NOV   2011

PRICING THE FUTURE

ALSO BY GEORGE SZPIRO

*Kepler's Conjecture: How Some of the
Greatest Minds in History Helped Solve One of the
Oldest Math Problems in the World*

*The Secret Life of Numbers: 50 Easy Pieces on
How Mathematicians Work and Think*

*Poincaré's Prize: The Hundred-Year
Quest to Solve One of Math's Greatest Puzzles*

*A Mathematical Medley:
50 Easy Pieces on Mathematics*

*Numbers Rule: The Vexing Mathematics
of Democracy, from Plato to the Present*

Pricing the Future

Finance, Physics, and the 300-year Journey
to the Black-Scholes Equation

A STORY OF GENIUS AND DISCOVERY

George G. Szpiro

BASIC BOOKS
A MEMBER OF THE PERSEUS BOOKS GROUP
New York

Published by
Basic Books, A Member of the Perseus Books Group
387 Park Avenue South
New York, NY 10016

Books published by Basic Books are available at special discounts for bulk purchases in the United States by corporations, institutions, and other organizations. For more information, please contact the Special Markets Department at the Perseus Books Group, 2300 Chestnut Street, Suite 200, Philadelphia, PA 19103, or call (800) 810-4145, extension 5000, or e-mail special.markets@perseusbooks.com.

Designed by Timm Bryson

Library of Congress Cataloging-in-Publication Data
Szpiro, George, 1950–
  Pricing the future: finance, physics, and the 300-year journey to the Black-Scholes equation: a story of genius and discovery / George G. Szpiro.
    p. cm.
  Includes bibliographical references and index.
  ISBN 978-0-465-02248-9 (alk. paper) — ISBN 978-0-465-02815-3 (ebook) 1. Options (Finance)—Prices—Mathematical models. I. Title.
  HG6024.A3S97 2011
  332.64'53—dc23
                          2011025031

10 9 8 7 6 5 4 3 2 1

*It is my sincere hope that
by the time you read this,
Gilad Shalit is back with his family.*

CONTENTS

PREFACE

Options have been traded for hundreds of years, at least since the sixteenth century, when they were used to buy and sell commodities in Antwerp and Amsterdam. But nobody knew what the true value of an option really was. For centuries, their prices were determined by supply and demand, with investors estimating their value on the basis of gut feelings. Indeed, it was not even known what determines the value of an option, whether the current price of the underlying stock, commodity, or asset, the rate of interest, investors' attitudes toward risk, the time remaining until expiration of the option, and so on. However, options do have a mathematically precise value. The equation that gives the correct price was found by financial economists Fischer Black, Myron Scholes, and Robert Merton in 1973. Their discovery was considered a singular achievement, comparable to Newton's discovery of the laws of motion. Scholes and Merton were awarded the Nobel Prize in 1997. (Fischer Black had died two years earlier, at the age of 57.) However, disaster followed the Nobel Prize. The spectacular near bankruptcy of Long-Term Capital Management, the billion-dollar company that Scholes and Merton had helped found, proved that high academic achievements do not guarantee financial success.

Spanning the period from the middle of the seventeenth century until nearly today, this book traces the historical and intellectual developments that led to the options pricing formula. It describes the search for the elusive equation but emphasizes the personalities behind that search. Some of the people who appear are medical doctor Robert Brown (of Brownian motion fame), three French accountants and stockbrokers (Jules Regnault,

Henri Lefèvre, and Louis Bachelier), Albert Einstein, Wolfgang Döblin (a German Jewish soldier in the French army during World War II), MIT mathematician Norbert Wiener, Russian pioneer of probability theory Andrey Kolmogorov, Japanese mathematician Kiyoshi Itō, and American economist Paul Samuelson.

At this point in a preface, it is customary to thank those who helped in the preparation of the book. Here I must make an exception. One organization to which I can offer no thanks is the Institut des Actuaires in Paris. It is one of the very few places where the early volumes of the *Journal des Actuaires Français* are stored, which contain some articles pertinent to this book. Unfortunately, after I had been given the runaround for approximately half a year by an extraordinarily unhelpful secretary, it was only when my wife visited Paris that the articles could be copied. It took her all of fifteen minutes . . . apparently too much for an unwilling secretary.

My sincere thanks do go to Tim Bartlett, Adina Berk, and Collin Tracy of Basic Books for making the text much more readable than it was at the outset, and to Chrisona Schmidt for careful copyediting. My agent, Roger Williams of New England Publishing Associates, is always there when I need him. My wife, Fortunée, did much more, of course, than just photocopy articles in Paris. I am also grateful to Simon Benninga, Franck Jovanovich, Wolfgang Hafner, and Heinz Zimmermann for offering encouragement and reading parts of the manuscript.

I had lots of fun researching and writing this book, and I hope you, the reader, will enjoy it too.

JERUSALEM, JUNE 2011

INTRODUCTION

In June 1940, in a barn somewhere near the western front, a young man wearing a French army uniform burns a sheaf of papers filled with mathematical symbols and equations. He must move quickly; German troops are closing in on the French village where he is hiding. The soldier, the German-born son of a famous Jewish novelist, is determined not to be captured alive by the Nazis and not to let his scientific legacy fall into their hands. A few weeks earlier he had sent a manuscript containing a novel mathematical theory to Paris for safekeeping by the Académe des Sciences. Now he must destroy any evidence of his work.

The sad story of Wolfgang (a.k.a. Vincent) Döblin is only a small part of the narrative that will be related in this book. For three centuries, accountants, speculators, investors, and scientists endeavored to find the holy grail of financial markets, the equation that could be used to compute the true value of a certain financial instrument: the elusive options pricing formula.

Like most chronicles of intellectual breakthroughs, this is a story of relentlessly driven and innovative people. I will tell this story of the development of ideas through the lives of the protagonists—accountants and economists, physicists and chemists and mathematicians, academics and professional traders. After preludes in seventeenth- and eighteenth-century Amsterdam and Paris, the intellectual action began in nineteenth-century France. In the first half of the twentieth century, it moved all over Europe, then to Russia and then to Japan, before it finally reached its climax in the second half of the twentieth century in the United States.

Say you want to build a house in the suburbs and you find a plot for $100,000. Since you can't afford to buy it until next year, the seller is willing to reserve it for you—at next year's price. The price could double or it could drop by half. What should you do? If prices fall, you will profit, but if they rise you won't be able to afford the plot. Then the real estate broker has an idea. For a flat fee, she will assume the risk. If the price falls, you will pay the lower market price; if it rises, you will pay a maximum of $100,000, with the broker making up the difference. What a great idea! You will profit if the price falls, and you will not have to bear the risk of the price rising. The question is, How much should you pay the broker? How much does she demand?

Or think of a farmer who will need to buy fertilizer in six months. Unfortunately the price is volatile and a high price would significantly cut into his profits. He cannot afford the uncertainty. A middleman offers him the following deal: pay me a flat fee, and in six months I will sell you 300 pounds of fertilizer for 60 cents per pound, no matter what the actual price may be. The farmer agrees, a contract is drawn up, the fee is paid, and the waiting begins. Six months go by. When the time comes to execute the contract, the price of fertilizer has dropped to 55 cents per pound. The farmer buys what he needs on the open market and lets the contract expire. The middleman breathes a sigh of relief and happily pockets the fee. Chalking up the fee as an insurance premium against the risk of higher prices, the farmer is also happy. He got the fertilizer for a lower price. The crucial question, however, is, How much did the farmer have to pay the middleman?

The two tales exemplify the use of options—contracts that give the right, but do not entail the obligation, to buy or sell something, usually a good or a security, at a certain date at a certain price. Options may be considered the granddaddy of the financial derivatives that are all the rage nowadays. They have been traded since at least the seventeenth century, with their prices determined by the market—by supply and demand. But is there an intrinsic, true value for options? Yes, there is. The discovery of the options pricing formula was a breakthrough, both in the history of ideas and in the effort to understand financial markets.

Many historians of science rank the options pricing formula, as developed by Fischer Black, Myron Scholes, and Robert Merton, up there with Isaac Newton's Universal Law of Gravitation, at least in terms of a scientific discovery. In the seventeenth century, the notion of action at a distance—earth pulling an apple off the tree—required an enormous intellectual leap by Newton's contemporaries. A similar leap was required from economists in the twentieth century, when it turned out that the value of options does not depend on investors' attitude toward risk.

Who would have thought that drunken sailors staggering around in the street, the random motion of minute particles suspended in a liquid, or the diffusion of heat in an object would be the starting points in the description of price movement on the stock exchange? Such processes, which later became known as Brownian motion, were investigated in the early twentieth century mainly by biologists studying evolution, chemists and physicists studying diffusion—among them Albert Einstein and several other Nobel Prize winners—and one forlorn mathematician dabbling in the stock market.

Serious attempts to ascertain the true value of options started to make headway toward the middle of the twentieth century. Even so, for a long time, the prices at which options were traded were still based on hunches and rule of thumb. Options trading was put on a sound footing only after the heroes of our narrative—Fischer Black, Myron Scholes, and Robert Merton—developed the sought-after formula, thereby discovering that the volatility of the underlying stock plays a crucial role and that the investors' attitude toward risk plays none. The feat earned Merton and Scholes a Nobel Prize in 1997. (Fischer Black had died two years earlier.)

By placing a value on options, the pricing formula made financial markets more efficient. The success of the Chicago Board Options Exchange, which today tallies about 5 million contracts a day, is due to this scientific achievement. Nowadays, financial instruments, based in large part on the methodology developed by Black, Scholes. and Merton, allow traders to buy and sell risk like any other commodity. For a certain price, risk-averse individuals can unload part or all of the uncertainty contained in their portfolio to investors who are willing to assume it, thus producing a more efficient economy and positively impacting people's lives. By utilizing tools from mathematics and physics to compute that price, Black, Scholes, and Merton and their predecessors may be considered representatives of a new profession: the quants.

Understandably, the quants were not content to simply enjoy their intellectual pursuits; they also wanted to make money—a lot of money. And indeed many did. Jules Regnault, the self-taught broker's assistant, died a very rich man. Merton and Scholes made and lost a fortune, and the spectacular blowup of their firm nearly resulted in the first financial crisis touched off by quants. More would follow.* And so this story of brilliant, driven, innovative characters is also a story of what may happen when greed and hubris get the better of us.

---

* Indeed, this book was written before the debt crisis of Summer 2011. Hence, the risklessness of U.S. Treasury bills, to which I refer several times in this book, is no longer guaranteed.

ONE

Flowers and Spices

I N THE 1630S AN UNPRECEDENTED FRENZY OF BUYING AND
selling seized large parts of the population in Holland. People sold
all their belongings and even went into debt in order to invest in a
commodity that had no intrinsic value. When investors finally realized
this, the price of the commodity plummeted and many lost their assets.
To save those clamoring most loudly for protection from further disaster,
the government took action, thus worsening the crisis.

What was this article of trade that caught the fancy of investors, speculators, and fortune hunters? Businessmen, mainly from Spain, Portugal, England, and Holland, had been roaming the world looking for extravagant
merchandise to take to Europe. Among these luxury items was a very special
flower found in Persia and Turkey. First described around the middle of
the sixteenth century by the Austrian ambassador to the court of Suleiman
the Magnificent in the Ottoman Empire—he described them as having little or no scent—the flower became popular among the European upper
classes, especially in Holland. It was the tulip. Admired for its variations in
color and beauty, the tulip soon became a symbol of wealth and opulence.

It was not wild tulips that elicited the Dutch people's passion but culti-
vated ones, many of them carrying a virus that gave their petals highly unusual patterns but also made them hard to grow and consequently all the
more rare and desirable. By 1636 this admiration for tulips went far beyond
an appreciation of their beauty. Tulip bulbs became objects of speculation,
first among the cognoscenti, then among simple folks, with investors buying them not for the esthetic value of the eventual flowers but in the hope
of rising prices. The journalist Charles Mackay devoted a chapter of his
1841 book *Extraordinary Popular Delusions and the Madness of Crowds* to
what he called tulipomania. Even though his description was not correct
in every detail, the name stuck.

Since no man of fortune could be without a collection of tulips, prepos-
terous prices were paid for a single bulb. "A trader at Haarlem was known to
pay one-half of his fortune for a single root, not with the design of selling it
again at a profit, but to keep it in his own conservatory for the admiration
of his acquaintance," Mackay recounted. One of the most celebrated species,
the Semper Augustus, whose petals exhibited dramatic red and white
streaks, was valued as the equivalent of twelve acres of building ground, and,
another time, as the equivalent of a carriage with two horses, harnesses, and
a sizable amount of money. No more than a handful of specimens existed
in all of Holland. The owners had acquired them with the specific obligation
not to cultivate them in order to keep their numbers low.

Anecdotes abounded, for example, the one about the sailor who con-
sumed a bulb of Semper Augustus for breakfast, thinking it was an onion,
and then spent a few months in jail on felony charges brought against him
by the owner. Or the British botanist who used his penknife to peel an unknown bulb that he found in the conservatory of a wealthy Dutchman and
then proceeded to cut it into pieces only to be informed by the furious
owner that he had just destroyed an Admiral van der Eyck.

The activities took on a frenzied pace. Markets were established on the
stock exchanges of the larger cities to facilitate the trade in rare tulips. In
smaller towns where no stock exchange existed, taverns became the meeting places for tulip enthusiasts, with the rich and the poor rubbing shoulders to trade bulbs. "Nobles, citizens, farmers, mechanics, sea-men,

footmen, maid-servants, even chimney-sweeps and old clothes-women dabbled in tulips," according to Mackay.

Tulip bulbs must be planted by the late summer, and the flowers bloom for no more than a week or two during the following April or May. The original bulb disappears but another one, and possibly some buds, appear in its place. These can be dug up starting in June but must be replanted by September. Hence actual trades could only be made during the four summer months. But the dealers did not take a break during the remaining eight months. Throughout the rest of the year, buyers and sellers made contracts with the intention of carrying them out during the summer. The traders agreed on the species of the tulip, the number or weight of the bulbs, the price, and the date of delivery and payment. Today such an agreement is known as a forward contract: the actual delivery of and payment for the bulbs are delayed to some specified date in the future. (A contract for immediate execution is called a spot trade.)

A code of laws was drawn up for the guidance of the traders, florists organized themselves into a self-regulating guild, and specialized tulip notaries were appointed. In spite of such protective measures, trading activities soon took on forms of gambling. Many homeowners converted all their property into cash with the sole purpose of purchasing tulips. Houses and real estate were offered at ruinously low prices so the proceeds could be invested in bulbs.

People had started buying bulbs in order to resell them at a profit, and prices spiraled ever upward to new heights. In early 1637 some single tulip bulbs sold for more than ten times the annual income of a skilled craftsman. Eventually sober-minded traders began to realize that these flowers were just . . . flowers. In February 1637 the bubble burst, and businessmen were left sitting on worthless bulbs. In Mackay's words, "Many who, for a brief season, had emerged from the humbler walks of life, were cast back into their original obscurity. Substantial merchants were reduced almost to beggary, and many a representative of a noble line saw the fortunes of his house ruined beyond redemption."

Once the bulb bubble burst, buyers refused to take possession of, and pay for, the useless merchandise. Day after day, defaults were reported from all over Holland. Sellers tried to get legal remedy through the judicial system, but the courts offered no help. Arguing that monetary disputes which resulted from tulip contracts were nothing more than gambling debts, the judges refused to get involved.

Tulip traders vociferously aired their grievances at public meetings. (The lucky few who had got rid of their stock in time invested the proceeds in overseas funds and kept mum, so as not to arouse the envy of their neighbors.) Representatives from all over the country were sent to consult with the elders in Amsterdam, the country's, indeed Europe's, center of commerce and trade. They came up with a good idea: let the buyers and sellers devise a plan. Discontent grew stronger. Loudest among those clamoring for relief were tulip buyers, including numerous state officials who had been making a pretty penny on the side and now faced the danger of losing it all. They wanted to renege on their obligations to pay for worthless merchandise. The traders turned to the Provincial Council in The Hague. After three months of deliberations, they came up with the age-old excuse: they needed more information in order to make a decision.

Eventually, on February 24, 1637, the guild of Dutch florists issued a decree. First, all tulip contracts made before November 30, 1636, were null and void. Second, all purchasers who had entered into a contract between November 30, 1636, and spring 1637 would be freed from their engagement if they paid the vendor a penalty of between 3.5 and 10 percent of the agreed-upon price. This was an unprecedented upheaval of traditional business practices. Whereas previously, buyers had been obliged to buy the bulbs they had contracted for, they now had a choice. If the price dropped between signing of the contract and the date of delivery, buyers had the option of simply walking away from the transaction. All they had to do was pay the vendor a relatively small premium. Thus traditional forward contracts were turned into what came to be known as options contracts. In due course, the decree was ratified by the Dutch parliament.

Predictably, this did not make everyone happy. Vendors would receive at most 10 percent of the sum they had contracted for, while purchasers complained that even that was too high a price for worthless onions. But the new regulation also created fresh opportunities. Since the risk of a transaction was limited to a fraction of the contracted sum, many businessmen saw prospects for profits, and the market for options contracts took off. If prices had been sky-high before, they now rose into the stratosphere.

Finally, at the end of February, the Dutch authorities put a stop to the trading. Prices plummeted. "Those who were unlucky enough to have had stores of tulips on hand at the time of the sudden reaction were left to bear their ruin as philosophically as they could," Mackay recounts. The country as a whole suffered from the aftereffects of tulipomania. "The commerce of the country suffered a severe shock, from which it was many years ere it recovered." The episode would be chalked up as a lesson—not always learned—for generations to come about not only the vagaries of luxury items but also the dangers that lurk in the trading of options.

While tulipomania, one of the most spectacular early aberrations of a market, can be ascribed to human stupidity, the Dutch East India Company failure in 1798 was due to a combination of bad luck, corruption, and incompetence. The history of this company is closely linked to the emergence of the first formal market for options.

It all started with spices. Throughout the ages, people have tried to put more flavor into the often bland diets they managed to scrape together from their gathering and hunting activities. Salt, for example, added some taste to food but was expensive. According to the naturalist Pliny the Elder, who lived in the first century AD in the Roman Empire, "the soldier's pay was originally salt, and the word salarium [salary] derives from it." Exotic seasonings, which were also used as medicines and preserving agents, were even more expensive. In the Middle Ages spices such as cinnamon, black pepper, cumin, nutmeg, ginger, saffron, and cloves were considered among the most luxurious items in Europe. Only the wealthy and the mighty could afford to season their opulent meals with them. Like silk, textiles, ivory, precious stones, and drugs, spices were imported from faraway lands by adventurous seafarers. Their locations of origin, often obscure islands, were usually kept secret by traders. The expeditions were expensive not only because boats and crews had to be outfitted but also because many ships did not return from their voyages. Shipwrecks were common and their costs, human and financial, had to be taken into account.

In spite of the dangers, the trade in spices blossomed. It was a welcome change from the dull dealings in salt, herring, wheat, and wine with their narrow profit margins. Lured by the prospect of great earnings, boats set sail for India, Java, Borneo, and China, braving the dangerous route around the Cape of Good Hope. Seafaring nations like Portugal and Spain effectively monopolized trade with the East. But traders from other countries, mainly the Netherlands, wanted to elbow their way into the lucrative business and began to finance the dangerous undertakings. The Iberians did not take kindly to the new competitors. To keep challengers from infringing on their profitable sinecure, Portugal and Spain closed their ports to foreign ships. Shortages of luxury goods ensued in the Netherlands and the price of spices rose.

Nevertheless, the Dutch persisted. They realized that boats equipped, staffed, and financed by individual businessmen stood little chance of completing their voyages. So, in 1602, seventeen businessmen founded the Verenigde Nederlandsche Geoctroyeerde Oostindischen Compagnie (VOC, United Netherlands Chartered East India Company), known as the Dutch East India Company for short. The company was organized around six "chambers," Amsterdam being the most important.

Pooling resources and sharing profits and risk was nothing new, in principle. For millennia, overland and overseas trips had been organized by consortia of businessmen who pooled their resources and shared the eventual profits or losses. Partnerships were formed ad hoc for individual voyages, with capital returned and profits distributed at the end of the voyage. VOC was different. It was to be a long-term venture, initially for twenty-one years. The charter was renewed several times, for a total of two centuries. Furthermore, participation was not limited to businessmen. Without being required to help run the company, ordinary citizens could buy shares and watch their capital grow . . . or diminish, all the while receiving dividends . . . or not.

VOC was led by seventeen directors, the Heren XVII, who made all the decisions: how many ships left from which port, where they sailed, what cargo they carried, and what wares they would bring back. The directors decided how to sell the cargo and whether to reinvest the proceeds in further voyages or distribute them as dividends to the regular stockholders, the *participanten*. The latter had no say in how the company's affairs were run. In fact, they did not even have the right to inspect the books. Their only function was to receive dividends whenever the directors decided to distribute any. If they were not happy with how the company was run or if they were disappointed with low dividends, they could return the shares and demand their capital back, or at least they could in the beginning. A few years later, the clause that allowed shares to be returned in exchange for a full refund was rescinded. As I will recount below, this would have important consequences.

The Dutch government was only too happy to oblige the company. After all, everybody would profit from increased wealth and higher tax receipts. The government gave the company monopoly powers and wide privileges. Nobody except the VOC was allowed to send ships from the Netherlands to the area east of the Cape of Good Hope and west of the Straits of Magellan. Neither was anybody but VOC allowed to trade. VOC's charter even conferred on it the right to wage war in the overseas territories—in what is today Indonesia and beyond—if the chief stockholders deemed it necessary in order to safeguard the company's interests. VOC became a state within a state. It owned 150 trading vessels and had 20,000 seamen and 50,000 civilians in its service. Significantly, it also had 40 warships and 10,000 soldiers under its command.

Its trading routes connected Amsterdam to Africa, India, the Persian Gulf, Japan, and China. But VOC did not limit itself to trade between the Orient and Europe—textiles from India, tea from China, pepper from Indonesia, silver from Japan, coffee from Arabia and Java, and, of course, fine spices from all over. It was also heavily involved in the Persian Gulf,

Zanzibar, Ceylon, India, China, Japan, and the islands of southeastern Asia. VOC traded spices for salt, salt for cloves, cloves for gold, gold for tea, tea for silk, silk for copper, copper for spices, and so on.

These traders did not go about their business quietly. VOC operations were very noisy, constantly accompanied by the clanking of war. The company acquired monopoly status by fire and sword: brutal military operations, massacres of local populations, murders of European competitors. The invasion and occupation of the Banda archipelago, a group of tiny islands north of Australia, gave the company a monopoly over nutmeg and mace, the conquest of Ceylon over cinnamon, the capture of Makassar over cloves. To further consolidate control over the spice trade, VOC envoys uprooted trees in many fertile areas in order to concentrate a valued crop on a single island in their possession. As a further blow to erstwhile competitors, VOC captured ports that had been used by seafarers as way stations on their voyages from and to Europe. By then, Portuguese, Spanish, and English traders had fallen far behind the Dutch.

In spite of the high costs of trading and waging war, VOC made huge profits. On average, goods were sold at VOC's semiannual auctions for nearly three times what they cost at their origin. And this was only an average. Rare spices often fetched three hundred times their purchase price. Even after the cost of building and outfitting ships was deducted, and after losses and shipwrecks were taken into account, dividends averaged over 16 percent annually for the first fifty years of the company's operations. VOC's good fortune was felt throughout the Netherlands. Magnificent buildings and the paintings of Rembrandt, Vermeer, and others attest to the wealth of the country.

When the company was formed, over one thousand individuals purchased shares in Amsterdam alone. They were a cross-section of the citizenry, with some investing as little as 20 guilders and others up to 85,000. Demand for shares was so great that even before they were actually issued, their price was already 10 to 15 percent above par. After all, it did not hurt to subscribe. By statute, shares could be returned in exchange for a full refund of the paid-in capital.

In 1609, however, the directors decreed that capital was no longer re-fundable. Whoever bought shares was stuck with them, for better or worse. But to have one's money tied up in the company's shares for many years could create problems. Shareholders might not be satisfied with the size of dividends or with the manner in which the company was run and would prefer to get rid of the share. Or a sudden illness, a home renovation, or a daughter's marriage might require additional funds. How could a business-man make use of his wealth if his money was tied up in VOC?

A so-called secondary market was created to give stockholders the op-portunity to liquidate their holdings. This was a stock exchange where shares of VOC could be bought and sold. Thus VOC in Amsterdam was the first company to issue stock that could be freely traded. The price of the shares fluctuated widely, depending on the fortunes or misfortunes, real or imagined, of the latest voyage. Astute traders quickly learned that they could influence the price of the stock by rumors of war or peace, of shipwrecks or safe voyages, of market gluts or shortages. In the first few years of its existence, about 6 to 7 percent of the capital of the Amsterdam chamber changed hands annually. By 1607 only two-thirds of the original owners remained.

Even foreigners were welcome to purchase a stake in the company. Whenever an exchange was made, the shares were registered under the names of the new purchasers in the company's stock ledger. But registration could only take place when the books of the company were opened, usually to record the payment of dividends. To allow a market to operate not only on these rare dates but every day, there evolved a market for so-called for-ward trades at the Amsterdam stock exchange. Whenever a buyer and a seller reached agreement, the "futures contract" was recorded but the actual transaction—the delivery of the title and the registration of the new owner—occurred on the specified date in the future.

Forward trades were carried out to eliminate the risk of future changes in the price of shares. Since price and date of the transaction are specified when the contract is signed, all relevant parameters are known at the outset and there is no uncertainty. But an element of risk remains. If the share price rises during that period, the seller will regret his decision because he could have received more money had he waited. If the price falls, the buyer will kick himself since he could have obtained the share for less money had he waited. In that sense, we are actually speaking more of regret than of risk. (There was also the risk that one of the parties would default on its obligation, but this could be remedied by taking recourse to the well-functioning court system.)

Apart from allowing stockholders to liquidate their holdings of VOC shares, the stock exchange had a further beneficial effect on the economy of the Netherlands. Entrepreneurs routinely needed to borrow money in order to finance ongoing operations and expand their business. To guarantee the loan, merchants had to put up their businesses, homes, or other assets as collateral. The interest rate on the loan was then determined by how secure the collateral was and by how readily it could be converted into cash in case the borrower defaulted. The shares of VOC were highly valued as collateral. The company was well-known and the ownership of the shares was clear due to registration in the company's stock ledger. But above all, the shares could be easily sold on the Amsterdam stock exchange. Thus there was less risk to the lender that he would be stuck with collateral that could not be converted into cash. As a consequence, interest rates fell, benefiting commerce in general.

VOC itself took advantage of this possibility. At the company's founding, total capital was about 6.5 million guilders. It stayed at that level throughout the company's existence; no additional capital was ever raised. So how did VOC finance its ever growing business? It borrowed money. Short-term capital needs were satisfied by issuing bonds for up to twelve months. VOC pledged its own assets as collateral.

However, for the lender there remained a question: what would the value of the shares be a year from now when the loan matured? If a businessman took out a loan corresponding to 90 percent of the share's value, but the price of the share fell by 20 percent in the course of that year, the lender would be in a bad spot. After all, the borrower would have every reason to default. Instead of returning the loan he could just walk away from his debt, leaving the lender with a share that was worth less than the amount he owed. The borrower would lose only his good reputation. Lenders would have to safeguard themselves against such a scenario.

And this is how, not long after the tulip fiasco, the market for options developed.

A lender needed to protect the value of the collateral. Afraid that the price of VOC shares would drop, he would look for an investor who believed that it would not. Then, for a fee, the two of them would enter into a new kind of arrangement. Say that the current price of a share of VOC is 100 guilders. Let's call the person who lent the money and is afraid that the share price will fall Pessimist, and the investor who believes the share price will not fall, Optimist.

Pessimist pays Optimist a fee of 3 guilders. In exchange Optimist undertakes to buy one VOC share from Pessimist one year from now, at the current price of 100 guilders, if the latter so desires. Other than paying the fee, Pessimist is not obliged to do anything. But when the day arrives and if the share price is, say, only 90 guilders, Pessimist will demand that Optimist buy the share from him for 100 guilders. Even if Pessimist is not in possession of the share (maybe the borrower did not default and was not forced to hand over the collateral), there is no problem. Pessimist simply buys it on the market for 90 guilders and delivers it to unhappy Optimist, who has to fork over 100. Altogether Optimist will have lost 7 guilders— he initially received the fee of 3 guilders but then lost 10 on the purchase. Pessimist, on the other hand, will have received the 100 guilders that he was hoping for at the beginning of the year, all for a fee of 3 guilders.

Let us now assume that VOC shares rise to 110 guilders during the course of the year. Come the day of reckoning, Pessimist will certainly not sell the 110-guilder share for the contracted 100 guilders. (He probably does not even have it because the original borrower would surely have repaid his debt and kept the much more valuable share.) The option will simply expire. Optimist will be happy: he gets to keep the fee of 3 guilders.

In modern parlance, such a contract is called a put option. I will say a bit more about put options below, and much more about all kinds of options in future chapters. Suffice it to say for now that options are a sort of insurance. For a small premium, the buyer of the put option buys the right to sell the underlying share at a specified date in the future at a specified price, even if the value of the share has fallen in the meantime. But he is not obliged to do so. It the price rises, he will let the option expire. The seller, on the other hand, is obliged to purchase the underlying share at the previously agreed price, if the buyer desires it.

Does this sound confusing? It did to Dutch speculators in the seventeenth century. This is why the merchant Joseph de la Vega decided to publish *Confusion of Confusions* in 1688, a booklet that attempts to explain the intricacies of the stock exchange. The full title was *Confusión de confusiones: Diálogos curiosos entre un philosopho agudo, un mercader discreto, y un accionista erudito, descriviendo el negocio de las acciones, su origen, su ethimologia, su realidad, su juego, y su enredo.* (Confusion of confusions: Curious dialogues between a keen philosopher, a prudent merchant, and a learned shareholder, depicting the stock business and its origins, etymology, reality, gambling, and plot). It may seem a bit strange that a speculator trading shares in Amsterdam would publish a how-to book about the stock exchange in Spanish, the language of VOC's bitter rivals. To explain the reason for this, I must recount a bit about Joseph de la Vega, the first person to clarify options in writing.

Joseph de la Vega was born in about 1650 to a Marrano family. Marranos were Jews who had been forced to convert to Christianity by the inquisitional forces in Spain. In contrast to most of their persecuted brethren who were killed or expelled, Marranos stayed in Spain, pretending to be Christians but clandestinely keeping their Jewish faith. They followed their traditions in secret, celebrating the Sabbath, for example, by lighting candles in a cupboard every Friday night. Joseph's birthplace is not certain; he may have been born in the Spanish city of Espejo whence his father hailed, or in Amsterdam, the family's ultimate destination after they fled the Inquisition. His parents were Isaac Penso Felix and Esther de la Vega. His mother's family hailed from a prominent Jewish family who had founded the Talmudic school in Livorno, Italy. In later years, Joseph would use both their names, Penso and de la Vega, interchangeably.

In spite of all the hardships, his father remained committed to his Jewish faith. While imprisoned in a Spanish dungeon, he vowed that he would return to Judaism as soon as he was set free and could leave the inhospitable country. And indeed, upon his release from prison, he fled to Antwerp and promptly returned to the fold. Isaac eventually became a wealthy banker, but he never forgot his unfortunate past. He became a benefactor, giving tithes of his income to good causes and to the poor. It is said that by the time he died in 1683, he had donated 80,000 guilders (about 500 times the yearly wage of a skilled laborer) for philanthropic purposes.

Isaac and Esther had five sons, the second of whom, Joseph, published his first literary work, a three-act drama in Hebrew, at age seventeen. Meant as an inspiration to Marrano youth, *Asirei ha-Tikvah* (The captives of hope) features a young man expounding on the high value of a virtuous life, holding up a mirror to his peers who had become accustomed to Spanish decadence in the years before their emigration. Subsequently de la Vega turned to business and became a wealthy merchant. But he never lost his taste for literature and wrote a large number of poems, moral and philosophical reflections, eulogies on princes, and novels. The novels, especially, became quite popular in their time.

The work for which he is remembered most today concerned the stock market. As he stated in the preface, he did not write the book merely for his own pleasure. On the one hand, his aim was to describe "this most honest and most useful of all businesses in Europe" to those who were not in the financial business. On the other hand, he wanted to disclose the tricks and treacheries that are employed by less than honest men, both in order to entertain and to warn.

In the spirit of Socrates and Plato, the setting de la Vega chose for *Confusion of Confusions* was a collection of four dialogues—or rather trialogues—between a merchant, a philosopher, and a speculator. While the doubtful philosopher and the skeptical merchant ask profound questions, the speculator extols on the beauties of the stock market, all the while explaining its intricacies. He describes what stocks are, how they are bought and sold, various forms of transactions, speculative maneuvers, the operations of the Dutch trading companies. With his book de la Vega meant to put a more respectable face on the scorned profession of speculation.

The first dialogue deals with the origins of the securities trade, the parties involved, and the various types of transactions. In the second dialogue, the author turns to stock prices and the reasons for their variation, bull and bear markets, the principles of speculation, the role that expectations play in the formation of prices, and the irrational behavior of market participants. In the third dialogue, the author focuses on stock exchange practices, as well as various types of transactions and operations and the activities and behavior of the brokers. The final dialogue examines the motivations and often dubious practices of *liefhebberen* (de la Vega uses the Dutch word for speculators who expect a rise in the stock prices—the word actually means lover) and *contraminores* (speculators who expect a fall).

De la Vega holds *contraminores* in low esteem, since they hope the companies whose securities they hold will experience bad fortune. And even if their hopes do not influence fate—who knows?—their prayers for shipwrecks, wars, mutinies, or bad business deals may be answered. Not surprisingly, *contraminores* are described as negative, unpatriotic wheeler-dealers. But *liefhebberen* too come under fire. Both types of operators often engage in unsavory practices: whispering incorrect information in an accomplice's ear while making sure that traders standing about can hear what is being said, dropping a letter to the floor with supposedly secret information, pretending to be a *contraminor* while actually being a *liefhebber*, among others. The fourth dialogue is replete with tricks that speculators use in their attempts to influence the market in a preferred direction, and the countertricks that a consortium of traders may use to defy such maneuvers.

De la Vega also describes dilemmas, such as when a broker who owns shares in a company is asked by a client to throw his client's shares onto the market. Does he follow the order in good faith, knowing that the value of his own shares will decline if a sudden glut occurs? To illustrate the mechanics of stock trading to the merchant and the philosopher, de la Vega also tells some real-life stories. For example, a *liefhebber* comes to a coffeehouse and pretends to have important news about a company. When asked about the price at which shares are currently traded, he adds a few percentage points and gratuitously announces that he has received many buy orders. At this point investors, believing that the share price has risen and will rise further, hurry to their own brokers and tell them to purchase the shares. In the meantime the first broker runs back to the stock exchange and arrives just in time to hear the other brokers place their buy orders. He is able to sell them his own shares at an exaggerated price, thereby making a bundle for himself.

Many of de la Vega's recommendations are still valid today: do not give advice to anybody whether to buy or sell shares; take your profits without regretting missed opportunities; do not get married to your shares; to win at this game one needs money and patience; the expectations of an event affects stock prices more than the event itself. Many modern investors would do well to take one recommendation especially to heart: "If Fortuna extends her hand and you have speculated well, accept it with humbleness and praise your luck. Do not spoil the stroke of fate through arrogance."

The stock exchange can serve as a playground for everyone, de la Vega writes. Hardly immune to bigotry and prejudice, he maintains that men of every profession, any character, and any nationality can let off steam at the market. Philosophers discover ferocious behavior, mathematicians find irrational figures, astrologers discern their lucky star. Poets sharpen their imagination, lawyers their sophistry. Barbers believe they can lather everyone in soap, seafarers find exciting adventures, soldiers a channel for their cunning. Painters seek perspective, spurned lovers forget their unhappiness. Gentlemen exercise themselves in patience, peasants in rudeness. Frenchmen find an outlet for their passions, Englishmen for their quick temper, the Spanish for their curses. Turks make use of their loudness; the Italians take advantage of their hypocrisy, Dutchmen of their cold-bloodedness, Germans of their arrogance, and Poles of their boastfulness.

But *Confusion of Confusions* is by no means a shallow pamphlet. It delves deeply into the mechanics of stock markets. At the end of the first dialogue de la Vega talks about *opsies*, transactions in which "you have only limited risk, while profits may rise beyond all hopes and expectations." The word *opsies* derives from the Latin *optio,* he explains, which indicates that one party of the transaction alone decides whether or not to exercise his option. Only the other side has an obligation.

Then he gets down to the actual description of options transactions by way of an example. Let's assume that the share price of a Dutch company stands at 580 guilders. Its ships are due from India laden with profitable merchandise, and peace in Europe is on the horizon. With good reason, the investor believes that the price of the company's shares will rise. But he is cautious. His assessments could be wrong and he does not want to take the risk of a decline in share value. So he does an options deal. He finds a trader who is willing to accept a certain sum of money, which is called a premium, in exchange for the obligation to deliver the company's shares at 600 guilders a piece, at a certain date in the future. Once the two parties have agreed on the premium, the investor pays it and then waits until the date of settlement, hoping for a large profit. The most he can lose is the premium that he has already paid.

How is that? Well, come the day of settlement, one of two things may happen. The share price may fall below 600 guilders. In that case, the investor won't take delivery of the shares at a price of 600 guilders, since he would be paying more than the shares are worth. He is not obliged to do so because the obligation to deliver the shares rests on the other side; he himself is not obliged to accept them. Thus he lets the option expire and the trader gets to keep the premium. On the other hand, the share price may be above 600 guilders on the date of settlement. In that case the trader must deliver the shares in exchange for 600 guilders, even if he has to buy them on the stock exchange for much more. Upon receiving the shares, the investor can turn around and sell them on the stock exchange for the higher price. Save for the premium that he paid at the outset, anything beyond 600 guilders will then be the investor's profit. In modern parlance, the transaction in which a premium is paid by one party in exchange for the other party's obligation to deliver the share is known as a "call option."

De la Vega also explains transactions that are nowadays called "put options." Let's assume that an investor believes the share price of a company will fall. A war may be looming, he has heard news of a mysterious shipwreck, there is a surplus of cloves on the market, and the prices fetched by the spices have dropped. For these reasons he would like to sell a company share at today's price, with a delivery date sometime in the future. He figures that by receiving today's higher price and then being able to buy the share at the prospective lower price, he will get to pocket the difference. However, he too is concerned that his judgment may be wrong; the share could rise instead of fall. So he finds a trader who is willing to accept the payment of a premium today, in exchange for the obligation to take delivery of the share at a certain date in the future but at the price specified today.

Let's say that a price of 600 guilders has been specified. If the price on the settlement date is 580 guilders, the investor buys the share on the open market and delivers it to the counterparty in exchange for 600 guilders. The counterparty must accept the loss of 20 guilders (minus the premium that he received at the start). The investor, on the other hand, who bought the share on the cheap but sold it for 600, makes a profit of 20 guilders (minus the premium). On the other hand, if the share price on the specified date happens to be above 600 guilders, the investor just walks away. By paying the premium he purchased the right, but did not enter into any obligation, to buy the share on the open market and deliver it to the counterparty. Thus his loss is limited to the premium he has paid.

De la Vega was the first to describe in writing the workings of the stock market, in particular the trading of options. *Confusion of Confusions* became something of a best-seller and was quickly translated into English. The book is a treasure trove of nuggets about the stock market. For example, the tips and tricks he described helped the advisers of King William III raise capital in London to finance his campaign against his rival across the channel, King Louis XIV. Unfortunately, it is a very cumbersome read. Numerous plays on words and endless strings of anecdotes from the Bible,

Greek mythology, and contemporary literature, which are meant to drive home simple points, make the going tough. At times, the discussion seems to serve as a vehicle to showcase the author's erudition.

De la Vega's influence is felt even today. The Federation of European Security Exchanges "believes that de la Vega's observations were highly accurate and are still of the greatest relevance today." Since the year 2000, the federation sponsors an annual prize in de la Vega's name for "an outstanding research paper related to the securities markets in Europe."

Original copies of the book have become collectors' items. When Sotheby auctioned a copy in the mid-1980s, the auctioneer thought it would go for £6000. It went for three times that price.

VOC survived the mayhem of tulipomania unscathed and did very well during the first century of its existence and beyond. Many shareholders became rich men. In 1720, 120 years after the company's founding, its shares were traded at twelve times their initial value. But as the eighteenth century progressed, VOC's fortunes began to decline. In 1708 two English companies merged to become the United Company of Merchants of England Trading to the East Indies. Renamed the British East India Company a few years later, it soon challenged VOC's hegemony. Competition from France also increased. At the same time, internal struggles and unrest overseas reared their heads, and consequently the company's income and profits decreased. The fourth Anglo-Dutch war, which broke out in 1780, weakened the Netherlands, which had hitherto eclipsed England as a seafaring nation. This further damaged business prospects. As good as fate had been to VOC and its stockholders during the seventeenth century, during the eighteenth everything turned against the firm. To keep investors happy, the company embarked on a dangerous course: it paid more dividends than it could afford. Mismanagement and corruption plunged the company into further troubles. In 1781 VOC shares were worth only a quarter of what they had been at the initial offering and just 2 percent of the value at their apex.

As debt kept increasing, the company's position became untenable. On December 31, 1799, two centuries after its founding, VOC was dissolved. In the 198 years of its existence, VOC had sent 4,700 ships with nearly a million sailors, soldiers, and traders on voyages. It was a sad end for the once proud and glorious corporation. The Dutch government took over the company's liabilities as well as all its assets. The overseas holdings were to form the basis for the colonial power of the Netherlands.

While options contracts figured prominently in the tulipomania turmoil, agreements akin to options contracts had already been used in various forms before they became fashionable in Amsterdam. The book of Genesis tells the story of Jacob, who worked seven long years for Laban in order to obtain the right to marry Rachel. In principle he could have walked away from the deal at the end of the period, chalking up the seven years of labor as the premium to be paid for the right, but not the obligation, to marry Rachel. As the story goes, Laban defaulted on the contract by substituting his elder daughter Leah for the beautiful Rachel. Given Laban's devious character, it is doubtful whether he would have let Jacob walk away after seven or even fourteen years. In that sense, the contracts should be seen as futures contracts, rather than options, which obliged both Laban and Jacob.

Aristotle told a story about the Greek philosopher Thales of Miletus who lived in the sixth century BC. Detractors made fun of the thinker, claiming that he had become a philosopher only because he could not make it in the real world. Thales decided to prove them wrong. The wise man had foreseen in the stars that the coming year would be a bumper crop for olives. In the winter preceding the harvest, he deposited what little money he had to reserve the use, if he so wished, of all olive presses in the city. Come harvest time, all presses were in the philosopher's possession; the former wise guys had to pay whatever he demanded in order to process their olives. Nobody ever questioned Thales's astuteness again.

Throughout ancient times, Romans and Phoenicians used contracts similar to options in their shipping and trading activities. During the first millennium AD, Jewish traders carried their knowledge about finance and financial markets from Mesopotamia to Spain. Until the late fifteenth century, Jews prospered in Spain under Muslim rule. But then the Inquisition started and darkness befell the Jewish community. In 1492 the Catholic monarchs King Ferdinand of Aragon and Queen Isabella of Castile forced Jews to convert or face death. The luckier ones were expelled and moved to Portugal. But they were not welcome there either and five years later were expelled again.

With time, many Jews, like Joseph de la Vega's father, gravitated to the Netherlands, at that time a Spanish possession, which had become known for religious tolerance. As such it had become a refuge for persecuted Protestants, Anglicans, and Jews. The latter again brought not only their traditions but also their business practices and knowledge of financial matters to their new homeland. It was an environment in which de la Vega and his coreligionists prospered. But hypocrisy and bigotry are hard to get rid of, and trade guilds remained closed to Jews. They were not allowed to practice law either. As a result, many immigrants became physicians, diplomats, and, above all, merchants and traders. The activities of the traders benefited the host countries, contributing to the operation of an efficient stock market and a functioning economy. De la Vega's book was a case in point.

TWO

In the Beginning

O N SEPTEMBER 24, 1724, KING LOUIS XV DECREED THE creation of a stock exchange, the Paris Bourse. It was a last ditch effort by the monarch (or rather the fourteen-year-old monarch's advisers and handlers) to put France's badly stricken finances in order. Due to the extravagance of Louis's predecessor, his great-grandfather Louis XIV, and his own advisors' mismanagement, the country found itself burdened with a horrendous debt. Everything was in disarray. The failed attempts by the Scotsman John Law to clean up this mess rendered the situation even worse.[1]

In 1694 Law had narrowly escaped a death sentence in England for fighting and winning a duel. The dashing twenty-three-year-old had shot and killed his opponent, a duped lover. Subsequently he fled to Italy and supported himself mainly by gambling, before being thrown out of Venice and Genoa by the magistrates, who feared his bad influence on the youth of their cities. He was about to be declared an undesirable alien in Paris too, but was saved from deportation at the last moment by Philippe d'Orléans, the regent while Louis XV was still a minor, who was the ex-lover of a certain Madame de Tencin, whom Law was advising in financial matters.[2]

John Law had concocted a financial plan that was supposed to save the failing French economy. Its basic principles had solid foundations in economic theory. His first, and very reasonable, suggestion was to replace gold and silver coins, which were a bother to carry around, with paper money. Law had tried to have this idea adopted in Scotland, in France, and in Italy. But it was too radical a departure from accepted norms—people distrusted pieces of paper, preferring coins made of precious metals—and he could not get any ruler interested.

When Philippe d'Orléans became regent of France in 1715, Law saw another chance. Neither devaluations nor fines nor new taxes had been able to repair the damage wrought on the nation by Louis XIV. Law suggested more than just the creation of banknotes as legal tender which, by itself, would have done nothing to rehabilitate the French economy. He had more ambitious plans. Philippe saw him as the savior he had been hoping for.

In 1716 John and his brother William founded the bank Law & Cie, whose business it was to print paper money. It was the first attempt in Europe to replace metal coins with paper money as legal tender. The bank, forerunner of France's central bank, guaranteed that the notes would always be worth at least as much as the metal coins on which they were based. But while coins could be devalued by the stroke of a pen, the paper would, by the bank's pledge, always keep its original value. After a few days of trading, it turned out that the public valued Law's paper money at 1 percent above the worth of the coins. At the same time, the *billets d'état* that had been issued by the government as security for the nation's debt were devalued by 78 percent. No wonder there was a run on the paper money issued by Law's bank.

A year later, encouraged by the bank's initial success, Philippe was ready to give Law wider powers. Convinced that a nation's wealth depended on trade, Law proposed the creation of a company that would exploit the natural resources of Louisiana and Mississippi, then French colonies in the New World, which supposedly were brimming with precious metals. The company would be incorporated, and its shareholders, including the royal household, the nation, and its citizens, would enjoy immense riches. To render the corporation immune to competition, Law convinced Philippe to grant it exclusive rights to all trade between France and the Mississippi and Louisiana territories. As was the case for the Dutch East India Company, such "consolidation" of business activities was a necessary precondition for perilous ventures, since no corporation could shoulder the considerable risk of outfitting ships for a dangerous voyage and then seeing its profits go to a competitor. However, while perfectly legal, granting a monopoly to a corporation, with one man at its helm, opened the door to the danger of manipulation. Nevertheless, when Law promised Philippe d'Orléans the sky—company profits would pay the government's debts—the regent saw a way out of his country's predicament and quickly agreed. With the more astute members of parliament predicting disaster, the regent later added exclusive privileges for trade with the East Indies, China, and the South Sea to the Mississippi Company, which was subsequently renamed Company of the Indies. Due diligence was not yet a legal prerequisite when going public, and the Law brothers saw nothing wrong in promising returns of 120 percent annually on the shares of their corporation.

Now the frenzy really started. A public eager to buy stock in the company beleaguered Law's offices from morning to night. Popular enthusiasm remained so great—even after there were no longer any shares to go around—that 300,000 additional securities were issued in the fall of 1719. Even then the demand did not diminish and the value of the stock kept rising. Law's house in the Rue de Quincampoix was constantly besieged by hundreds of hopeful patrons, and even the real estate prices in his neighborhood moved northward. A cobbler whose stall adjoined John Law's house rented it out to brokers, thereby making more money than he had ever earned cobbling shoes. It was said that the population of Paris increased during that time by a quarter million people from all over Europe who converged onto the offices of Law & Cie.

While the people with money assembled to do business, throngs of spectators came to watch. Every evening, a troop of soldiers had to clear the street. The commotion became too bothersome for the neighbors and Law moved his business to the Place Vendôme which, in turn, soon came to resemble a country fair. Among all the wheeling and dealing, tradesmen sold refreshments and drinks, gamblers set up roulette tables, and pickpockets had the time of their lives.

Buying into Law's company became the favorite pastime of *tout Paris*. The agitation was so great and the noise so loud that the judge at the nearby courthouse complained to the municipality that he could no longer hear the advocates argue their cases. At the beginning of 1720, Law & Cie moved again, this time to the Hôtel de Soissons, which was purchased from the Prince de Carignan. The savvy prince retained ownership of the gardens around his former estate, however, and rented out the premises to hundreds of brokers, merchants, and peddlers who set up their businesses in tents and booths. Charging monthly fees of 500 livres for each stall, the prince raked in a quarter of a million livres every month.

The *haute volée* as well as the *hoi polloi* thronged to the Hôtel de Soissons and the surrounding tents and booths. A law had been enacted that made trade in the Mississippi and the East Indies stock illegal anywhere except at the grounds of the Hôtel. As wealth increased, so did crime, and robberies and assassinations were the order of the day. Since Protestants were barred from becoming public officials in France, John Law had himself baptized in order to be allowed to become Contrôlleur Général des Finances.

Prospective buyers tried every trick in the book to gain access to Monsieur Law and induce him to sell them shares in the companies. Hopeful visitors paid his servants handsome tips, just so they would announce their names in Law's presence. Noblemen who would have been insulted to be kept waiting by the regent for twenty minutes gladly waited six or more hours for an audience with Law. In the meantime, he may have been busy with trivialities, like writing a letter to his parents' gardener about planting cabbage or playing a hand of cards with a friend from back home. The ladies of society employed other devices to get his attention. One lady of note asked her coachman to drive her carriage into a lamppost just as she passed Law so that the gallant Scotsman would rush to her aid and she could place a buy order. Another society woman had a fire alarm sounded during a dinner that Law attended, so she could bump into him when he ran for safety. (He saw through the trick, however, because while everybody else was running out of the house, she was heading straight in.)

There are numerous stories of the rise from rags to riches. Law's coachman had become enormously wealthy and decided to acquire his own equipage. The first time his coach arrived, the former servant forgot who he was and proceeded to climb to the top of the vehicle where the coachman had his seat. Only after being reminded that he owned the carriage did he climb down again. A maid came into so much money that she bought fine gowns and jewelry. In the opera she was recognized by her previous mistress, who mocked her. The former maid rose in all her splendor and said, for everyone to hear, "Yes, Madame, I am Marie, the cook-maid; I have gained some money at the Hôtel de Soissons; I like to be well-dressed; I have bought some fine gowns, and I have paid for them. Can you say as much for yourself?"

Philippe d'Orléans was so enamored with the financial success that had taken place under his aegis that he took Law's scheme—creating money out of paper—one crucial step further. Seeing no need for the precious metals on which the paper's value was based, he decided to do away with them altogether. By late spring of 1720, the state had printed another billion livres. But the new money was not backed up by gold or silver, and the country became inundated with paper currency. Inevitably the bubble had to burst. The value of Law's banknotes and of his company's shares tumbled, and many, many Parisians lost their money and their fortunes. John Law, who had been Contrôlleur Général des Finances for five months, resigned in May 1720 and was hounded out of Paris. Happy to have escaped alive, he ended his days in Venice (whence he had been thrown out in his youth) in 1729. The shares of his trading company, issued at 500 livres in May 1719 and having risen to close to 10,000 in January 1720, bottomed out at 50 livres in March 1721.

––––––––––––

It was in this atmosphere that the advisers of the fourteen-year-old Louis XV realized that human motivations needed to be directed into orderly channels. Lust for trading and speculation did not cease after the John Law debacle. The new financial instruments that aroused intense public interest were special liquidation certificates called Visa. Actually, Visa was the agency in charge of canceling the national debt by issuing certificates in lieu of the debt instruments that John Law had left behind (banknotes, bank accounts, bonds, shares). Visa were meant to clear up the mess that the entrepreneurial, if irresponsible, Scotsman had left behind. Inevitably a lively and speculative market arose around the very security that was supposed to rein in excessive speculation. Prices for Visa certificates rose from 1,400 livres to over 2,500 within a month, before falling back to the original price within another few weeks. Soon unofficial trading could no longer be ignored. Newspapers reported prices as if the market were legal, speculators bought and sold frantically, and a new disaster began to loom. Moreover, the government itself was badly in need of a trustworthy and well-functioning market. The state had to borrow money but nobody would be willing to lend to it in exchange for bonds, if these securities, just pieces of paper after all, could not be traded, should the need arise. Citizens would be willing to buy government bonds only if a secondary market existed that would allow them to offload the stocks and bonds whenever they needed liquidity. A well-functioning stock exchange would allow the government to borrow money from the public by issuing bonds.

So Louis XV and his government decided to bring some order into the booming business. On September 24, 1724, the official bourse was created by royal decree. It was managed by the Compagnie des Agents de Change, a tightly organized, self-regulating monopoly of a few dozen brokers. They were the sole middlemen authorized to trade in government bonds and other listed securities. No need to be politically correct by saying middlepeople; women were not allowed to enter the trading hall. Since trading in stock "was found to encourage a passion for gambling among the gentler sex," a contemporary guidebook said, ladies "are not . . . generally allowed to enter during hours of business."[3]

In addition to being male, agents were required to be over twenty-five, of French nationality and of Catholic faith, with an impeccable reputation and a minimum of five years' experience in banking or commerce. Above all, they had to be men of means, since opening a brokerage at the bourse entailed substantial expense. A successful candidate had to buy an office and post a security bond, considerable outlays that became even greater as the brokerage business became more lucrative. With the enthusiastic backing of existing agents, the government kept the number of privileged brokers who were permitted to operate at the bourse artificially low. Their exact number varied somewhat over the years but never exceeded sixty.

These official brokers were organized in a corporation, the Compagnie des Agents de Change, which was, in turn, governed by a syndicate made up of six elected members. The agents made good money by charging a commission of 0.25 percent on each trade, paid in equal parts by buyers and sellers. The brokers were only permitted to execute clients' orders and were not allowed to trade on their own account. This prohibition was meant to level the playing field for regular customers. Like the modern ban on insider trading, this prohibition prevented brokers from using better and more timely information for their own profit. Had they been allowed to trade for themselves, outside investors would always have reason to be afraid that if a broker offered to sell him a security it would probably entail a loss. After all, the potential investor would reason, if there were even a whiff of a profit, the broker would keep the security for himself. With such mistrust, trade would soon have stopped completely.

In March 1774 the trading area was put on a raised platform dubbed the *parquet*, in reference to its floor covering. Brokers stood around a circular railing, the *corbeille*, and the area was itself enclosed by a fence. Nobody but the agents and their assistants were allowed into the area between the fence and the *corbeille*. Only clerks rushed in and out with the clients' orders. Once in possession of a note which indicated a customer's willingness to buy or sell at a certain price, the broker sought a counterpart around the *corbeille* willing to fill the order. The trick was to catch another broker's attention. Eye contact was everything, well, nearly everything. Other advantages were a loud voice and the ability to gesticulate wildly, anything to get noticed by an *agent de change* who had a buyer or a seller for one's securities. By the way, the loud voice was not only meant to make oneself heard. In fact, it was a specific requirement of the bourse that prices be announced *à la criée,* that is, shouted out loud. It was meant to ensure transparency of trading.

With much business and few agents to split the pie, brokering was a very worthwhile occupation in spite of the high entry price. The system worked all right until the French Revolution in 1789. Then everything changed. The revolutionaries, paragons of economic theory and thought, were bent on curbing privileges. Soon they cast a disapproving eye on the goings-on at the bourse and in 1791 ended the Compagnie's monopoly. They liquidated the agents' offices, albeit returning their security bonds. Henceforth anybody was allowed to trade in securities as long as he paid a tax and took an oath of loyalty to the constitution. Predictably, chaos ensued, bankruptcies multiplied, fraudsters and swindlers had a field day.

In the meantime, the revolutionary government badly needed money, but none was forthcoming, particularly because the revolutionaries themselves had instigated a tax strike in order to weaken the government. But then Charles-Maurice Talleyrand hit upon a good idea: nationalize all church property, sell it, and use the proceeds to run the revolutionary state. Talleyrand, who would later become foreign minister, knew what he was talking about; he himself had been a bishop before he abandoned the clergy. In November 1789, the National Assembly decided to put church property, valued at about 3 billion livres, at the state's disposal.

But selling property takes time, and the state coffers were nearly empty. To bridge themselves over, the revolutionaries decided to issue bonds and sell them to the public. They were called *assignats,* paid 3 percent interest per year, and were backed by the seized church property. Thus the treasury's funds would be replenished and the holders of the *assignats* would get their money back as soon as the church property was sold. Once redeemed, the *assignats* would be burned. But the revolutionaries had made a grave mistake. They had not counted on overenthusiastic officials who printed too many *assignats* and counterfeiters who flooded the market with fakes. England, desiring to weaken the revolutionary government, also swamped France with bogus securities. In addition, the redeemed *assignats* were not always burned but often found their way back into the market.

By 1796, 45 billion livres in *assignats* were circulating, backed by only 3 billion livres in property. As a consequence, inflation was rampant. The *assignats,* which were by now treated as regular banknotes, became nearly worthless. Draconian laws—prison sentences of up to twenty years for selling *assignats* at less than their face value—did not stem their depreciation. Most loans, unlinked to the rate of inflation, with fixed interest rates and long durations, also lost most of their value. Many borrowers grasped at the occasion to pay off their debts at bargain rates, much to the chagrin of creditors who received only a fraction of what they had initially lent. The damage to the trustworthiness of the credit market was immeasurable. In 1793 the bourse was closed and remained shut for four years, even after the bloody Reign of Terror had ended, during which 40,000 people were executed by the guillotine.

Soon the economic situation became untenable. In August 1795 the franc was declared legal tender, and in September 1797 the state reneged on two-thirds of its debts. This brought some calm to the situation but at the painful price of increased mistrust in the state's financial institutions. At the same time, demand for liquidity, the need for investments, and the lust for speculation never ceased. To relieve these pressures, the Paris Bourse was reopened in 1796 and a year later, everybody and his uncle were again permitted to participate in the unregulated stock exchange.

The most important securities that were traded at the new bourse were the *rentes,* perpetual government bonds on which interest is paid forever but whose principal is never repaid. Many rentiers, often older people, widows, and orphans, lived off the proceeds of their *rentes.* This was to the government's political advantage: since the *rentes* had no expiration date and were backed by the state's trustworthiness, the holders had a vested interest in the continued welfare of the government. No revolution had to be feared from these quarters.

Rentes could be bought and sold on stock markets with prices varying according to the people's confidence in the state and its institutions. When many people offer to sell their *rentes* in exchange for cash, indicating a lack of confidence, the price drops. Hence the price of *rentes* was an excellent barometer for the public's trust in the stability of the regime, and the powers that be watched the market price with riveted eyes.

With brokers and speculators operating in the unregulated bourse, the price of the *rente* varied all over the place, much to the displeasure and shame of the new strongman, the First Consul and later Emperor, Napoleon Bonaparte. The former artillery general had come to power in November 1799 when the government was bankrupt, the troops were unpaid, and no taxes were being collected. Napoleon founded the French central bank, issued paper money, and forced everybody to pay taxes, thus providing income to the state and ending inflation. Now it was time to show that confidence had been restored. He did so by housing the bourse in a new, specially commissioned building, the Palais Brongniart, and by reinstating—with decrees in 1802 and 1807—many of the commercial practices, regulations, and customs that had been in force before the revolution. The monopoly of the *agents de change* was reestablished and trading was meticulously organized and effected by not more than eighty authorized brokers who once again jealously guarded their privileges.

The agents had a monopoly over the trade in government bonds and other securities quoted on the bourse. For the privilege of working on the *parquet* and charging commissions on each trade, they had to post a bond with the government that would cover possible claims by wronged customers. Insider trading was not allowed. A broker could not make a deal between two of his own customers, for each trade, there had to be separate agents for the seller and the buyer of a security. Furthermore, each successful transaction had to be announced by shouting the price out loud. The trades were recorded in brokers' journals and published in the evening's or next morning's papers. These regulations were meant to ensure that potential customers were fully informed and that agents would enjoy the public's complete confidence.

One important restriction on the brokers' operations was that so-called futures contracts (a.k.a. short trades) remained outside the law. These financial instruments, which specify a price for a security and a date in the future when it would change hands, were not legal on the *parquet*. One reason was that the existence of a futures market would engender speculation, which is exactly what Napoleon wanted to avoid. Another reason was that futures contracts seemed in bad taste. After all, short selling means agreeing to deliver a security that one does not yet possess, in the hope that one will be able to purchase it at a lower price before the settlement date. The seller makes a gain if the security falls in the meantime. Hence, a short sale is tantamount to betting on, and hoping for, the issuing company's or the state's bad fortune.[4] No wonder, then, that upright businessmen would want such objectionable practices to be declared outside the law. Self-respecting agents would never get mixed up with such unsavory matters.

The futures market, the *marché à terme,* was not specifically illegal, but it was certainly frowned upon. To discourage forward contracts, brokers were required to be in possession of the selling client's securities and the buying client's cash before any trade was permitted. And if a futures trade was effected in spite of the ban and a disagreement arose, it was considered a gambling dispute. After all, no merchandise or cash had changed hands, so no suit could be filed and debts were unenforceable.

But an activity is in good or bad taste only in the eyes of the beholder, and tastes can and do change. And futures contracts are not all bad. A farmer, for example, never knows what the next harvest will bring, whether it will be a good year or a bad year. By short selling next year's crop to a willing speculator at a price agreed on today, the farmer assures a certain income for himself and avoids uncertainty. When the time comes, the much maligned speculator may make a killing or he may lose his shirt, but he does so willingly. At the same time, by entering into a futures contract the farmer assured his and his family's livelihood.

So it is no surprise that futures contracts were in high demand. Whenever there is a demand for illegal products or activities among willing adults—gambling, alcohol, drugs, prostitution—the activity simply moves underground. In the case of futures contracts, it just moved outside the bourse. Traders, willing to deal in forward contracts and other securities off-limits to the *parquet,* gathered in front of the bourse to do business. They were called "curb brokers" or *pieds humides* (wet feet), since they congregated in the street, no matter what the weather conditions. Today we would call this type of bazaar, which was known semi-officially in Paris as the *coulisse,* an over-the-counter market.

The *coulisse* competed fiercely with the *parquet.* Unhampered by customs and regulations, curb brokers were quicker and more efficient in picking up on commercial trends than their traditionalist counterparts inside the bourse. Compared to the *coulisse,* the *parquet* was the epitome of staid conservatism, in spite of the shouting and gesticulation around the *corbeille.* In addition to being innovative and savvy, curb brokers, open to competition from any comer, charged lower commissions. No wonder that the *parquet* did not look on them kindly. The agents' anger toward them often turned into open anti-Semitism since many *coulissiers,* barred from the "Catholic Frenchmen only" *parquet,* were Jews.

By the late nineteenth century much of securities market had slipped away from the *parquet* and toward the *coulisse.* The unofficial forward market was, by then, much larger than the official cash market and leading bankers petitioned the government to legalize the forward market. An imperial commission also recommended doing so, but the approaches were rebuffed and the futures market was left in limbo.

In an attempt to regain business, the *parquet* gradually began disregarding the requirement that securities and cash be at hand before a transaction could be made. As the volume of forward contracts increased, specific rules evolved around the market. The considerable risks that futures contracts entailed were kept to tolerable levels by requiring customers to deposit a certain amount of money with the brokerage. Should the security's price change more rapidly than expected, this amount would cover possible losses. When setting the so-called margin, the broker had to weigh costs and benefits. The margin had to be sufficiently high to prevent the risk of default, but not so high as to scare off customers. As long as price changes were within a normal range, the customary margins prevented defaults. But if—through a broker's inattention or a customer's unwillingness to cough up additional funds—price changes climbed beyond the margin, then, come settlement day, the concerned broker had to absorb the loss. With a total of only sixty brokers to trade with each other, it was very probable that one agent's loss would affect many of his colleagues. If one of them defaulted due to inordinate losses, a disastrous chain reaction could follow.

And this is what eventually happened. In 1882 the Société de l'Union Générale, an investment bank with a strong and unsavory political agenda, went bust and pulled the bourse with it. Let's consider the developments in detail. In 1878 Paul-Eugène Bontoux, an engineer who had previously worked for the Rothschild railroad company, decided to found a bank in order to help with investments in industry, commerce, mining, transport, and construction, mainly in central and southern Europe. Aiming to challenge the Jewish-owned Rothschild empire, he hit upon a unique marketing strategy. Building on the anti-Semitism, xenophobia, and racism prevalent in large parts of French society, he appealed to ultranationalist, conservative Catholic investors for business. Many former aristocrats, clergymen, and royalists followed the call and became his clients.

Bontoux induced investors to buy the shares his bank issued—mainly for building railroads in the Balkans—to deposit their savings in, and take loans from, his bank. It was a booming market, the bank's securities rose and rose, and frenzied speculators jumped into the forward market.[5] Investors continually renewed their forward contracts in the expectation of steadily rising prices, while brokers happily pocketed commissions from the increased volume. Many speculators, like gamblers bitten by the bug, went heavily into debt in order to finance their habit, thus exposing themselves to even more risk. Dozens of new securities were issued each month and sold like hotcakes, and more financial institutions were created with the specific mandate of giving out loans to speculators. On settlement days, the fifteenth of each month, the exchange was overwhelmed with paperwork; brokers and clerks labored through long nights to clear the trades. It was the John Law affair all over again.

Then one day the bubble burst. The disaster was precipitated when the Austro-Hungarian government refused to license the creation of a bank. On January 4, 1882, news of the refusal reached Lyon, France's second largest city, and the shares of the Crédit Lyonnais tumbled, losing two-thirds of their value in nine days. Investors in the forward market who had expected these securities to keep on climbing were faced with ruin. Unable to make good on the forward contracts of their clients, the brokers' liabilities mounted and the Bourse of Lyon was forced to cease operations. Belatedly, and with hindsight, the district attorney remarked, "For the last six months, Lyon has been seized by feverish speculation. Outrageous fortunes were made within days. It appears that the stockbrokers forgot their professional duties and deliberately abetted behavior that was nothing else than pure gambling."

Then catastrophe hit the Société de l'Union Générale. As soon as the problems began at the Bourse of Lyon, financial irregularities in the Union Générale's books became apparent. It had less capital than it reported, traded in its own stock, falsified its quarterly report, and showed fictitious profits. On January 5, the price of its shares had stood at 3,040 francs. With rumors about its unsavory accounting practices beginning to circulate and the settlement date looming, the share price dropped to 2,800 on January 14. Two days later it fell to 2,400 and on January 18 to 1,400. On January 30, Bentoux was forced to shut down the Union Générale. The share price of his house of cards, a proud empire just four weeks ago, had dwindled to 600. He declared bankruptcy a few days later.

Many speculators who faced ruin, men of honor and high social standing, did not hesitate to take the easy, if dishonorable, way out of their predicament: they simply walked away from their forward commitments. Pleading gambling debts that could not be legally enforced, they left their brokers to fend for themselves. Bontoux also sought a sleazy way out. He blamed his institution's troubles on Jewish bankers and freemasons, and claimed that he had been the victim of a syndicate that had conspired to depress his bank's shares and ruin him. The conspiracy theory found credence in certain French circles but did him no good in court. He was sentenced to five years in prison for accounting fraud. However, just before his arrest, he managed to flee France.

The aftermath of the crash was nothing less than a disaster. A recession hit the French economy that lasted for several years. Bontoux's creditors came out of the affair with no more than a black eye. Since the Union Générale had many real assets like railroads, about 80 percent of the company's debts were eventually repaid. The *agents de change*, on the other hand, fared far worse. The bourse in Paris avoided total collapse only with the help of an emergency loan granted by the Bank of France. (The same was not the case for the Bourse of Lyon, which goes to show that being physically close to the center of power is definitely an advantage.) Seven of the sixty brokerages in Paris went bankrupt, and in Lyon nearly half of the twenty licensed agents became insolvent.

Many contemporaries blamed the press for kindling the feverish interest in the forward market and then fanning the hysteria. However, responsibility must be put squarely on the doorsteps of the speculators, investors, and gamblers. On February 17, 1882, the *New York Times* stated that "the Union Générale was the most foolish and unbusiness-like establishment that figures in the annals of modern speculation" and went on to comment, with little sympathy for the losers, that much of "the money lost by the latest holders of the stock . . . has been lost in imagination only. Certain fortunes have disappeared which were real, but most of them have been illusionary." With barely concealed glee the newspaper quotes the French proverb "Ce qui vient par la flûte s'en va par le tambour." (What comes by the flute, disappears by the drum, or easy come easy go.) The affair inspired the great French novelist, playwright, and political journalist Émile Zola to write the novel *L'Argent* (Money) in 1891, in which a French speculator is ruined by a Jewish banker.[6]

One positive consequence of the affair was the legalization of the forward market in 1885. Henceforth the *marché à terme* would be strictly regulated and futures contracts legally enforced. Speculators could no longer claim gambling debts and walk away from their obligations.

Note: In August 2011, two centuries after Napoleon banned short sales at the Paris Bourse, Belgium, France, Italy, and Spain again instituted that measure to calm violent markets.

THREE

From Rags to Riches

T HE FRENCHMAN JULES REGNAULT, BORN IN BÉTHEN-
court in 1834, is one of the forgotten heroes of our story.
Starting out as a lowly broker's assistant at the Paris Bourse,
he was not content to simply buy and sell stock in the hope of making a
profit here and there. He wanted to understand why some investments
were profitable while others went awry. An astute observer, he quickly
gained a deep understanding of the financial market. That his theories
were sound, if inexact, is attested to by the fact that he died a rich man.
However, his most important legacy was not the considerable fortune that
he left behind, but an insightful book in which he explained his findings.

Regnault was the first person to try to understand the workings of the
stock exchange in mathematical terms, and his explanations had all the trap-
pings of a scientific (if unrigorous) theory. Titled *Calcul des chances et philosophie de la bourse* (Calculation of probabilities and the philosophy of the stock
exchange), his book appeared in 1863, nearly two centuries after Joseph de
la Vega wrote *Confusion of Confusions*. How many people actually read it is

unknown, but it must have been noted by his colleagues. It is nothing less than amazing that a lowly *agent de change* with, at best, an incomplete university education should have produced such a thoughtful and profound work. Who was Jules Regnault?

His full name was Jules Augustin Frédèric Regnault. An older sister had died before he was born but he had a brother, Odilon, seven years his senior, and eventually a younger sister, Delphine. The father, a customs official, died nearly penniless when Jules was twelve, and the family moved to Brussels. As the son of a destitute family, the nineteen-year-old Odilon was excused from his military obligations and helped make ends meet by working as a scribe. At the same time he audited courses in higher mathematics at the Free University of Brussels, having been also exempted from paying tuition. But he never graduated. His brother Odilon also profited from a free high school education. Painstaking searches through all available university records in Brussels have found no documentation, however, that Jules Regnault ever studied there.[1]

Whatever education Jules Regnault received was certainly inspired by the Belgian scientist Alphonse Quételet. A polymath active in mathematics, astronomy, meteorology, demography, sociology, and criminology, Quételet was the first to apply statistics to the social sciences. Until then, its use had been restricted to astronomy where it helped quantify and rectify observation errors. For example, the concept of normal distribution only referred to the distribution of errors in the measurement of celestial orbits. Quételet applied it to the measurements of human characteristics. After that, it was only a small additional step to apply the normal distribution to the measurement of profits and losses in financial markets.

When Jules was about twenty-eight years old, he and his brother Odilon moved to Paris. They shared a tiny apartment, which, by virtue of its size and location just below the roof, allowed them to avoid paying taxes. Their modest dwelling had another advantage: it was close to the Paris Bourse, where starting in 1862 they worked as broker's assistants. They had arrived at a time when several years of steady growth were about to lead into a period of high volatility, thus creating both risks and opportunities for investors.

The instability led to debates among professionals in finance, law, and economics, as well as among politicians, on how perturbations in the market can be avoided, growth sustained, and calamities shunned. Bearing in mind both the aberration of tulipomania and the initial successes of the Dutch East India Company, some of them concluded that the key to understanding the meanderings of the bourse was to differentiate between financial investments, which are indispensable to an economy, and speculation, which was considered a reprehensible aberration driven by avarice and greed. So how could speculators be prevented from disturbing the orderly functioning of financial markets, thereby holding up economic growth? In order to answer questions such as this one Regnault wrote his book.

Working as broker's assistants, the brothers managed to improve their situation and after a few years of hard work both of them obtained their own living quarters. Odilon married but passed away a year and a half later. We can only guess that, left alone, Jules threw himself into his work with even more passion.

Shortly after arriving in Paris, Regnault began work on *Calcul des chances et philosophie de la Bourse*. How this self-taught broker's assistant had acquired—in only a couple of years—the intellectual tools to pen a treatise that anticipated future developments by dozens of years remains a mystery. Probably Jules sat down to work on his 50,000-word essay in the evenings, after a full day's work at the bourse, in the small room under the attic that he shared with his brother. A year after he started as an assistant at the stock exchange, he had developed—all by himself—a theory of financial markets that presaged work that would win the Nobel Prize in physics and in economics in the following century.

His work garnered limited interest. It probably made the rounds among Regnault's colleagues at the Paris stock exchange, but apart from them and some actuaries, only a few scholars used, or at least cited, it. Among them was John Maynard Keynes, the foremost economist of the early twentieth century. It is doubtful, however, that he actually read it. In any case, *Calcul des chances et philosophie de la Bourse* was considered sufficiently important for the Bibliothèque Nationale de France, the British Library, the University of Toronto, and the Catholic University of Leuven to acquire copies, and even the Library of Congress procured one, albeit twenty years after it first appeared, classifying it under the rubric "finance/speculation."

The quote Renault chose for the book's frontispiece, from the Bible's Book of Wisdom, revealed his guiding principal: "O quam bonus et suavis est, Domine, spiritus tuus in omnibus! . . . omnia in mensura, et numero et pondere disposuisti!" (O how good and sweet is thy Spirit, O Lord, in all things! . . . Thou hast ordered all things in measure, and number, and weight.) Regnault believed that "the movements on the stock exchange are subject to universal laws of attraction, in the same manner as the earth when it describes its path around the sun." Hence, "under certain conditions, an investor can be assured of profits, as surely as the seasons [of the year] return." The clause "under certain conditions" is crucial, because Regnault maintains that profits do not loom for short-run investors. They are assured of bankruptcy just as surely as the seasons return. But we are getting ahead of ourselves.

Regnault considered short-term speculation morally distasteful—as did many of his contemporaries—because the investors' sole motive was the lust for quick profits. The activities of financiers who constantly buy and immediately sell stock were one of the reasons for the volatility that then reigned at the bourse. Long-term investments, on the other hand, are beneficial because they provide the funds needed for the economy's development and growth. Regnault reserved the term "speculation" for long-term investments; a short-term investment was called a gamble or a game (*jeu*).

In his book, Regnault uses basic facts about probability theory to explain how stock prices behave. To state the obvious at the outset: prices can move up (*hausse*) or down (*baisse*). The movement is linked to events that determine the fortunes of a company or of the economy as a whole. With good news, prices are likely to rise or with bad news, fall. So far so good. But one of Regnault's crucial insights was that the event does not need to actually occur in order for the stock price to react. The expectation of news, good or bad, suffices to cause a change in the price.

With this, Regnault recognized the crucial role of information in the determination of prices in financial markets. He maintained that the current price of the stock contains all information, even about future developments. "Nous savons que ces conséquences, si elles existent, sont contenues dans le cours actuel," he wrote, speaking of the consequences of events that would happen only far into the future. ("We know that the consequences, if they exist, are contained in the current price.") Regnault anticipated one of the most important facts of financial theory, a basic principle that would be precisely formulated only a century later: the current price incorporates all information, including expectations about future developments.

Thus when news about a certain event is announced, the price of the affected stock will seek a new level. There will be oscillations at the outset, as a result of the initial surprise and indecision among the investors, but eventually the price will settle down. Regnault noted, however, that the prices of stocks reflect more than just the raw information. They also incorporate investors' hopes and fears and different interpretations about the expected effect that the event will have on the company's performance. Hence differing opinions necessarily lead to different assessments of what the correct prices should be. Even though investors—about five to six thousand operated on the Paris Bourse in the second half of the nineteenth century—possess identical information, their estimates of the correct price level differ. Regnault compares the situation to a scenario of many people estimating the height of a building. There is one true value, and the average of the people's guesses will approximate the correct height. But the individual estimates will be dispersed, distributed in a symmetric fashion on either side of the true height.[2]

And a good thing that is too, because if everybody were to agree on the correct value of a share or a bond, financial markets would not exist. Buyers would find no willing sellers if everybody interpreted all information in the same way. It is precisely the differences in the assessments that make trade possible. One investor believes the price of a security will rise and seeks to buy it, while another believes the opposite and seeks to sell.

Once securities have reached their true prices, the residual fluctuations between *hausse* and *baisse* are nothing more than a game to Regnault . . . and a boring one at that. The vacillations are purely random, he maintained, with equal chances of prices rising or falling. Hence after a stock has reached its true value, short-term movements are similar to the tosses of a coin. If this were not so, if the chances of either a rise or a fall were greater than 50 percent, everybody would soon find out and the price would quickly adapt to the general assessment, rising or falling accordingly. Hence, were it not for the speculators' belief that their ability and experience give them an edge over the randomness of the coin tosses, trading on the bourse, like tossing coins, would lose all interest.

How many lucky winners, Regnault lamented, boastfully ascribe their success to wise decisions while in reality their triumph was nothing more than the fortunate outcome of random events? And how many smart people who fail in the stock exchange could have achieved prosperity in other employment? Even the best-written treatise on how to play on the bourse, the smartest trading strategy, the wisest theory will not guarantee a single cent in profits, Regnault wrote, as surely as there is no method to consistently win at coin tosses.

There are only two assured ways of making money on the bourse through short-term trading, he continues. One is what we now call "insider trading," buying or selling stock based on information that is unknown to the public. In most countries, such a practice is illegal nowadays. If investors had to be afraid that tricksters with inside knowledge could take them for a ride whenever they bought or sold shares, they would stop investing. Eventually stock exchanges would disappear. But with laws in place to level the playing field, investors can be certain that they know as much about a firm's situation as the next man, and everybody has an equal chance of making a profit. In nineteenth-century France, however, insider trading was not illegal but was considered ungentlemanly. Regnault had harsh words for this unsavory practice: "Tricher n'est pas jouer" (swindling is not playing).

The other way of making a certain profit on the market, according to Regnault, is available only to big operators. Their trades can be so huge that they influence the direction of the market. This is not necessarily because of the trades themselves, Regnault remarks, since each purchase is offset by a sale at a price corresponding to the true value of the security. Rather, as soon as an investor large enough to send quivers through the market makes a move, a mob of small-time players will blindly follow. As a consequence, demand for securities may outstrip supply, driving the price up, or vice versa.

Finally, Regnault mentions a ruse that is so far beyond the pale that he does not count it even among the less than honest money-making schemes. It consists in announcing a rumor whose veracity cannot be verified immediately but which the originator knows to be untrue. The buzz will cause general panic or euphoria among the traders, leading to precipitous sales or purchases below or above the true value of the securities. When the truth becomes apparent the next day, it will be too late for the hoodwinked traders. Honest investors would blush just thinking about such a ploy.

So can an honest short-term investor at least stay even? No, he cannot, is Regnault's sobering answer. Even though *hausses* and *baisses* are distributed fifty-fifty, players on the bourse cannot hope to offset their losses with their gains. In the long run they will always lose, he maintained. Worse, the inevitable outcome of prolonged trading on the stock market is eventual bankruptcy. The reason for this sad prospect are transaction costs.

Brokers at the bourse in Paris, as everywhere, were entitled to a commission on each trade. The amount the *agents de change* demanded seemed insignificant. When Regnault was writing his book, the *courtage* charged in Paris amounted to one-eighth of 1 percent of the total sum traded. But these small charges quickly added up with the frequency of the trades. This effect was magnified, as we will see, when investors speculate with money that they obtained from their brokers on credit.

By the time Zola published his novel *L'Argent,* this sad fact had become common knowledge. Speaking about a certain Capitaine Chave, Zola wrote that "his big argument against playing on the stock exchange was that mathematically the player always had to lose. If he won he had to deduct the courtage and the stamp tax. If he lost he had to pay these amounts on top of his losses. Thus, even if he won as often as he lost, he still had to pay from his own pocket the stamp tax and the commissions."

Regnault tried to dampen the impact of the painful message. "How could you, dear players, even hope that the game, repeated indefinitely, could end without your ruin?" he wrote in *Calcul des chances*. "Even if you manage to double or triple your fortune, you will never have finished and the mysterious, invisible adversary—your counterpart, the bourse—must be considered as possessing an infinitely large fortune. So, let it be known that even if the chances [of gaining or losing] are strictly equal, you always have the ABSOLUTE CERTAINTY of being ruined." Some consolation!

He even predicted the precise moment when the investor would go bankrupt. If he put up 1,000 francs as security, and the broker allowed him sufficient credit to speculate with 20,000, the investor would pay a *courtage* of 25 francs on each trade. Even though this was just one-eighth of 1 percent of the traded sum, it actually amounted to 2.5 percent of the player's initial investment. With price increases and decreases approximately equal, Regnault calculated that the capital would be eaten up by the commissions after about forty trades. Hence, if the player was in the habit of liquidating positions two days after investing in them, he would lose all his money in about three months.

Regnault anticipated future concepts like the all-important notion of the perfect market, which maintains that all publicly available information is contained in the current price of a security. He also realized the lure and the danger of insider trading. Then he made an even more startling discovery. Using the bourse's historical data for the years 1825 to 1862, he explored the relationship between profits that were made on the bourse and the time that stocks are held. Regnault correctly spoke of deviations (*écarts*), not profits, because the price could move up or down. He analyzed the results by computing averages, and the first thing he noticed was that the deviations tended to be equal for equal time periods. "Les écarts sont égaux pour des temps égaux." Further, he found that as the time period between purchasing a stock and liquidating it was lengthened, the deviation between the prices tended to become larger. This does not sound unreasonable, and certainly not very surprising, since a price has more occasion to move a large distance when the time period is long.

But Regnault discovered something much more amazing: whenever the period during which the stock is held is doubled, the deviations in price grow by about 1.41, when the time period is tripled, the deviations grow by about 1.73, and if the holding period is quadrupled, the price deviations approximately double. A regularity in the stock market? Now this was something to ponder. To a numbers buff, the implication is clear: 1.41, 1.73, and 2 are the square roots of 2, 3, and 4. Regnault could hardly contain his excitement; he had hit upon a new law of nature. "For the first time we here formulate a mathematical law which nobody had ever expected until now: the difference between the prices of a stock is proportional to the square root of elapsed time."

Having made his momentous discovery, however, Regnault was hard-pressed for an explanation. He did produce one and it was ingenious, if naive. In order to illustrate his explanation, Regnault took recourse to the mathematical concept of imaginary numbers. While regular numbers can be represented on a straight line and can move up or down, so-called imaginary numbers are represented on a plane and can also move left and right. Extending this notion to security prices, Regnault envisaged the concept of "imaginary prices" that can move not only up or down but in any direction in the plane. Even though erroneous, it was a brave attempt and attests to Regnault's analytic powers. His argument went like this: envisage a security's true price as a point in the plane. Within a certain time period, its price can deviate by a certain distance away from this point. The maximal distance that the price can move is unknown, and since deviations can occur in all directions, the final landing point will lie somewhere within a circle. The outer limit of this circle, and hence its radius, indicates the maximal distance that the price can move within a certain time period.

Hence, his thinking went, the price of the security can move around in any convoluted manner within the observed time period, ending up anywhere within the confines of the circle. Thus any point on the whole surface of the circle can be reached. Then, with a leap of faith, Regnault concluded that the surface of the circle represents the elapsed time. Since a circle's radius is proportional to the square root of its surface (remember that the circle's surface is equal to π times the squared radius), the maximal deviation of a price should be proportional to the square root of time. Regnault's argument was far from rigorous and not quite correct. For example, he confuses average price movements with maximal price movements and, of course, his identification of the circle's surface with time is cute but unwarranted. Nevertheless, his observation, if not its justification, is path-breakingly accurate, as we will see later: stock prices do move in proportion to the square root of time.

But if the square root of time law is universal, why is it that the distance traveled by bodies falling under the force of gravity[3] or the lengths of oscillating pendulums[4] are in proportion to the square—and not the square root—of time? Simple, says Regnault. A stock price moves from its equilibrium (the true share price) toward some point within a circle around that value. A physical body, on the other hand, moves from its current state toward its equilibrium, which is represented by the center of a circle. Hence, Regnault's imaginative reasoning goes, oscillating bobs and falling bodies follow the inverse of the law of stock prices; thus distances are proportional to the square of time.

To Regnault the discovery meant no end of astonishment and admiration in the workings of Nature—or God. "The variations of the stock exchange are subject to mathematical laws! Events that are the result of the whims of man, of quite unforeseen political blows, of intricate financial combinations, of a multitude of occurrences which have nothing to do with each other, all combine into an astonishing ensemble. Chance is nothing but a word, empty of any meaning," Regnault exclaims. Then, resuming the motto on the frontispiece of the book, he sends a warning: "Now learn and be humble, Princes on Earth who proudly dream of holding the destiny of the people in your hands, Kings of Finance, who control the riches of the State, you are nothing but frail and docile instruments in the hands of the One who—like the Bible says—has measured, counted, weighed and distributed everything in a perfect order." He finishes his tribute to the higher order of things with the dictum "L'homme s'agite, Dieu le mène." (Man moves but God leads him.)

Regnault ends the first part of his book lamenting the short-term speculator's fate. "He has studied everything, he has foreseen everything, he has a profound knowledge of the possibilities and the subtleties of speculation." But all this is of no help to him, Regnault sighs compassionately. "*Hélas!* Why does he not realize that his rapid ruin, as certain as the revolutions of the planets around their orbits, is the inevitable consequence of probabilities and their combinations?"

And with these sad words Regnault turns to the second part of his book, which offers a more optimistic perspective: useful speculation. In contrast to *agiotage,* the term he uses to designate the reprehensible chase after quick profits, which, as often as not, turns out to be in vain anyway, *la speculation utile* affords the opportunity for real gains, benefits the country's economy, and is a praiseworthy enterprise.

Regnault does not mince words when describing the moral inferiority of "improper investing," even before he defines what he means by that. *Agiotage* serves solely to satisfy greed, the lowest instinct of mankind. Relying on ignorance and lured by the prospect of quick profits, he writes, *agioteurs* employ detestable maneuvers to exploit and abuse the workings of the bourse. Since by their actions they exaggerate any deviation of the share price from its true value, they destabilize the market. Parasitic and animal-like, their dirty dealings are a disgrace to honest, productive economic behavior. On top of everything else, such improper investing is futile. Regnault compares it to a night out with the guys, playing games of chance like "la rouge et la noire." The next morning, game over, the players will possess more or less what they started with the previous evening. A redistribution of wealth may have taken place but, on the whole, a night spent at "the green carpet" is a totally unproductive exercise. And if all that were not enough, Regnault now reminds the reader of the consequences of the inevitable transaction costs. In contrast to gambling sprees with buddies, where one player's loss is the other player's gain, the bourse demands its due both from winners and losers. As he pointed out in the first part of his book, after playing the bourse for a while, the cumulative effect of *courtage* (transaction costs) will unfailingly entail the *agioteur*'s ruin and dishonor. Serves him right!

In contrast to these miserable dealings, useful investments represent the best of everything. Not only do they guarantee certain, if modest, returns, they also have the ability to create, build, and increase wealth for the common good. Such useful and honest speculation does not expect a fortune in a single day, earned without labor and without pain, but makes do with modest gains, accumulated over long periods of time. With the bourse as intermediary, honest speculation benefits the economy, correcting exaggerated price movements that may have been brought about by blind confidence or mindless panic and providing the funds needed to ensure a functioning system. In short, he writes, useful investments cannot be sufficiently lauded and should be encouraged by all governments.

But the distinction between wholesome speculation and gambling, Regnault admits, often is quite thin. After all, when does faith turn into superstition, praiseworthy thriftiness into abject stinginess, careful spending into extravagant prodigality? When does the act of eating turn into gluttony, the act of drinking into drunkenness? To answer that question, Regnault delves into a discussion about cash transactions and futures contracts.[5]

Cash transactions are the plain vanilla kind of stock market transaction. Seller and buyer meet and agree to trade securities. They settle the deal on the spot, with the buyer taking possession of the securities and the seller pocketing the money. Futures transactions are a bit more virtual. They constitute an agreement between a seller and a buyer to trade stock at a specified price at a specified date in the future. At the time the contract is made, the seller need not possess the securities and the buyer need not come up with the cash. Only when the agreed-upon date arrives will the deal be consummated. Then the seller must procure the securities on the open market, at whatever price he can, and deliver them to the buyer who then pays the previously agreed-upon sum. In modern parlance, such contracts are often called short sales.

Many contemporaries tended to associate futures trading with gambling. The idea of selling something that one does not possess does have an objectionable tinge about it. Dealing in futures contracts was technically illegal in France until 1885, even though it had been widely practiced and implicitly recognized by French jurisprudence since 1860. But Regnault disagreed. Many futures contracts are impeccably serious, he maintains, while some cash transactions are nothing but reckless gambles. So, if dealing in futures contracts in and of itself does not constitute improper investment, what does? Maybe an investment becomes improper if, when the contract is made, the buyer does not own the financial means to buy the share or pay the seller? No, says Regnault, even such a practice would not necessarily count as improper, as long as the parties can guarantee delivery of, or payment for, the shares when the date of settlement arrives. If they are solvent, if they will have sufficient funds to fulfill their obligations, they pose no danger to the orderly operation of stock exchanges and may be considered legitimate investors.

The real scum, Regnault believed, are investors who are overdrawn, who do not, and will not, have the means to carry out their commitments, either because they do not have enough money to fulfill their obligations or because forward contracts were written in numbers larger than the available shares. Their guilt is compounded if the frequency with which these investors perform their misdeeds is high. If they proceed with their unconscionable activities day in, day out, the bourse may find itself in real danger.[6] And this is where, in Regnault's opinion, the line between decent speculation and shameful *agiotage* is crossed.

Think of what happens when the day to settle the futures contracts arrives. The seller must procure the shares but, unable to purchase them, turns to another willing seller who, himself an overdrawn speculator with no hope of acquiring them, turns to a third, and so on. When it becomes apparent that the chase for the elusive shares is futile, the price rises toward infinity. At the same time, potential buyers realize that this particular share is not available and conclude that it is unreasonable to insist any further. Now, without demand, the share's price drops to zero. So the price would simultaneously be zero and infinite, a situation that obviously cannot occur. No wonder Regnault chastised "improper" investors so adamantly. They destroy his beloved market.

But not to worry; transaction costs will teach those dubious operators who play the stock market day after day a painful lesson. With an average interest rate of 3 percent, which the investor could earn if he holds on to his investment for a year, and with brokerage fees of one-eighth of 1 percent, Regnault calculated that it would take only two transactions a month to wipe out all profits. The foregone income becomes all the more painful if, on top of the direct *courtage* paid to the broker, the time and mental effort expended by the investor are included in the transaction costs.

The second part of Regnault's book is more than just a hymn to proper, and a denouncement of improper, speculation. It is a theoretical discussion of long-term investment. Regnault was convinced that in the long run, the prices of securities are deterministic—determined by economic factors—and, like the orbits of the planets, can be predicted. His main interest in the remainder of his book was to determine the true price of a certain bond, the "3 percent *rente.*"

A security's price is influenced by two components, Regnault maintains, accidental causes and constant causes. Since accidental causes are, well, accidental, their effect is transitory and Regnault does not consider them any further. Instead he concentrates on the constant—the long-term causes. In principle, he suggests, the price of any good should be determined by the amount of work that went into manufacturing it. But this is only an approximation, since many other factors play a role in determining the price of a good or the true value of a security. Whatever these factors, however, in the final analysis prices are determined through negotiations between sellers and buyers, that is, by supply and demand. Surprisingly, Regnault then states without further justification that theoretically the quantity demanded must grow in inverse proportion to the square of the price. However, this is true neither in theory nor in practice, and it is not clear how Regnault obtained this erroneous result. He himself notes that there are many reasons why his purported "principle" does not hold in practice and considers it no further.

Of all economic variables that influence the price of a share or a bond, Regnault singles out the return—the rate of interest that the investor receives—as the most important long-term factor. He then takes the main financial instrument of his time, the famous *rente perpetuelle* 3 percent, as a test case and subjects it to close scrutiny. This government bond, issued by the French government in 1825 at a price of 100 francs in order to indemnify noblemen and others who had lost their properties during the revolution, paid the holder 1.50 francs twice a year—hence the "3 percent" in the bond's name—even though the interest could be far from 3 percent, as we shall see presently. The government committed itself to paying the interest until eternity, and in principle the principal would never be repaid. If the government wanted to retire its debt, it would have to re-purchase the outstanding *rentes*.

So were onetime investors stuck with the commercial paper for the rest of their lives (and the lives of their heirs)? Not at all! A market developed around the *rentes*, and whenever the holder of such a bond was in need of cash, he could simply sell it to a willing buyer. There was no guarantee of the price, however, and the *rente's* price would vary according to the public's confidence in the government's ability and willingness to continue to pay the interest. If confidence was high, demand for *rentes* would also be high and the price would rise; if the public's confidence diminished, investors would try to offload their holdings and the price would fall.

Since the interest payments were fixed at 3 francs per year while the price of the *rente* varied, the de facto interest rate also fluctuated. It was 3 percent only for as long as the price stayed at 100 francs. But of course the prices varied over a broad range. Examining the daily data from the thirty-eight-year period between 1825 and 1862 (more than 11,000 data points!), Regnault noted that the *rente's* prices fluctuated between a low of 32.50 and a high of 86.65. Hence the interest varied between 9.2 percent (3 francs/32.50 francs) and 3.5 percent (3 francs/86.65 francs).

Regnault realized that interest is the most important factor in determining the price of a share or bond. He asserted, this time correctly, that—according to one of the most fundamental laws of economics—the interest rate represents the risk that the investor incurs. If the shares of an enterprise are twice as risky as the *rente* (i.e., if the firm's risk of defaulting on its interest payments is twice as high as that of the French government), its bonds should return twice the interest. However, one should not go for the highest-paying bond. Diversification is the name of the game. "Never put all your eggs in one basket," Regnault cautions, drawing on an adage that was old even in his time.[7] Like businessmen who distribute wares to be shipped onto several boats, lest one of them sink, Regnault advises investing in different firms simultaneously, so that a bankruptcy of one does not entail the loss of an entire fortune. Of course, there are limits to diversification, he admits, since trying to stay informed about every possible firm and every development would require a lot of time, and time is money.

Let us return to Regnault's estimation of the true value of the *rente* 3 percent. The midpoint between the highest and the lowest prices on the stock exchange during the observed period is 59.6 francs, which struck him as being far too low. To get a better grip on the numbers, Regnault inspected the 11,000 prices in his database more closely and discovered that price quotations were not uniformly distributed within the range of 32.50 to 86.65 francs. In fact, they were not even distributed symmetrically to the right and to the left of the computed midpoint of 59.6. Regnault next attempted to compute the midpoints between the highest and the lowest prices for each of the thirty-eight years separately, and then computed the mean. Thus he obtained 72.48, and if he had redone the exercise for each day he believed that he would have received a price close to 73 francs.

Regnault was not yet satisfied, however. The average value (*valeur moyenne*) that he had computed could be considered the "true" value of the *rente* only if there were an equal number of trades on each side of it, he suggested. In this case the *valeur moyenne* would be equal to the *valeur probable*. In modern statistical parlance, we would say that the mean would be equal to the median.

Fortunately, Regnault came up with a roundabout procedure to compute the true value of the bond, even if the mean is not equal to the median. Taking the highest and the lowest price for each month, 900 data points for the thirty-eight-year period, he computes the probable error of the estimate "d'après une formule importante que nous ne pouvons qu'indiquer ici" (according to an important formula which we can here only state).[8] This probable error is 1.20 francs. Now come some arithmetical acrobatics. Since 72.48 lies about three-quarters of the way between 32.50 and 86.65, the extreme low and high values, Regnault adds three-quarters of 1.20 to the previous best estimate of 72.48, and thus sets his very best estimate for the true value of the *rente perpetuelle* 3 percent at 73.40 francs (=72.48 + 0.90). *Voilà!*

Once he was in possession of the holy grail, the *rente*'s true value, hidden to all except to the select few who had read his book, the rest was easy. Any deviation from the true price could be exploited for profit. Buy the *rente* when the price is below its true value, sell it when the price is above. Sell it short when the price must fall, buy it short when it must rise. Mind you, all this was meant only for long-term investments because transaction costs were bound to bankrupt short-term speculators. So, perform the investment and then hold on to it.

Regnault had ample opportunity to make use of his techniques. Knowledge of the *rente*'s alleged true price, coupled with his awareness based on painstaking statistical analysis that prices move on average with the square root of time, that the *rente*'s price always falls by 1.50 francs the day after the coupon has been paid,[9] or that seasonal variations (e.g., due to vacation periods) create deviations from the true price, allowed him to generate handsome profits.

For nearly two decades, Regnault traded on the bourse on his own account. By 1881, he had accumulated so much wealth that, at age 47, he no longer needed to work. Moving into ever larger apartments, he could afford to live off the interest of his financial holdings. He employed a housekeeper, a coachman, and a gardener, owned horses and three carriages, which he kept in a stable that also housed the coachman in comfortable quarters (with toilet!). He purchased a large piece of land in the resort town of Enghien-les-Bains, a dozen kilometers northeast of Paris, which he enlarged a few years later with the purchase of an adjoining plot. The house he built on his property served him as a summer residence. Henceforth, he spent his summers pursuing pastimes such as photography and fishing and boating on the lake of Enghien-les-Bains.

When Regnault died in December 1894, he left behind a considerable fortune valued at just over a million francs, comprising bonds, stocks, and real estate. In today's currency this would correspond to about $3.5 million, depending on how purchasing power is taken into account. Since Jules never married, the beneficiaries of his estate were his mother, his sister, some distant cousins, the gardener, to whom he bequeathed 2,000 francs, his coachman, to whom he left his horses and carriages, and the town of Enghien-les-Bains. To the latter he bestowed his summer home, which today houses the administrative offices of the town's famous casino, and the adjoining terrain, which was turned into a soccer field. In recognition of his benevolence, the town fathers named a street the Rue Jules Regnault. The remainder of his estate was to be divided among two orphanages and two other charities. His last will also specified that his grave in a previously purchased plot in the Père Lachaise cemetery in Paris would be marked only by a simple stone. The funeral was to be held according to the wishes of "my dear mother whose religious beliefs I respect," even though he himself would prefer it to be a civil ceremony. His preference for a nonreligious memorial service notwithstanding, Regnault did believe in God. In fact, divine laws were fundamental to his thinking about the workings of the bourse: God ruled not only the movements of planets in the solar system but also the movements of prices in financial markets.

Investors like Regnault not only got rich but also provided necessary funds to the government and to productive industrialists. His mathematical derivations may seem a bit muddled at times, but Regnault certainly had the correct intuition. Thus, even though we do not know for sure whether his book had much influence on future theoreticians, the fact that he died a rich man attests to the basic soundness of his theories. With the benefit of hindsight, it is clear that Regnault had written a truly path-breaking book. Modern financial theory was born in 1863.

FOUR

The Banker's Secretary

R EGNAULT WAS AMONG THE FIRST ECONOMISTS TO USE
mathematical and probabilistic formulations to describe financial phenomena. Until then, and even for some time afterward, social scientists disdained math, preferring verbose descriptions of
the phenomena they were trying to understand. Even Regnault had to
couch his arguments in prose to avoid alienating traditional economists.
But scientific rigor and conciseness are achieved through equations, and
the prevailing disregard for mathematical tools prevented a more precise
understanding of the workings of the stock exchange. So maybe pictures—one step short of full-fledged mathematics—would do the job?
The first person to introduce a graphical conception of economical
events, to view phenomena in a geometrical framework, was the accountant Henri Lefèvre, a sometime personal secretary to the banker of all
bankers, the überfinancier James de Rothschild. In contrast to Regnault,
who actively speculated on the bourse and enriched himself as a result,
Lefèvre remained an observer, albeit a keen one, and an educator. His
graphical elucidation of options trading remains the preferred method of

explaining it in textbooks on finance. Bowing to the wishes of the times, however, he too refused to bring mathematical symbolism into play.

Even though Lefèvre and Regnault lived at the same time in the same place, it is not known whether the two men met or even knew of each other. Regnault's senior by seven years, Lefèvre was born in 1827 in Châteaudun, a town situated about 100 kilometers southwest of Paris. In contrast to his younger colleague, Lefèvre obtained an education and earned a degree in the natural sciences in 1848. Unable to find a teaching position, he embarked on a career in economics, as an investment adviser and a financial affairs writer. Such a cushy combination of professions, patently illegal nowadays, was common practice in mid-nineteenth-century Paris. In fact, in 1881 no fewer than 228 Paris tabloids were devoted to financial matters, not counting the nearly 100 dailies that had their own financial pages. Scandals were inevitable, of course, with publications tempting naive speculators with the promise of large profits. Naturally, the real winners were the editors and publishers of the dubious newsletters and tip sheets.[1] This is an example of how norms change. Imagine the outrage today if journalists for the *Wall Street Journal* or the *Financial Times* were allowed to advise close confidants to trade in the shares of the companies they would extol or trash in next day's paper. In nineteenth-century France, there were no scruples. On the other hand, as we have seen, the trading in futures contracts, nowadays an established and important financial activity, was illegal until 1885.

Not to be outdone by any of his journalist-cum-financial adviser friends, Lefèvre edited a Spanish economic review published in Paris, *El Eco Hispano-Americano,* before founding the Comptoir Central de Souscription, a firm that effected transactions for small provincial investors who lacked access to the bourse in Paris. In 1869 he was one of the founders of the Agence Centrale de l'Union Financière, and—oh, yes—editor in chief of its press organ, the *Journal des Placements Financiers.* In 1873 Lefèvre became a full member of the French society of actuaries, le Cercle des Actuaires which had been founded two years earlier, and for a while he worked for l'Union, the most important insurance company in Paris.

Starting in the 1870s, Lefèvre also began publishing theoretical essays in the much more serious *Journal des Actuaires Français* (Journal of the

French Accountants), the Cercle's official publication. He wrote books on investment and speculation with titles like *The Art of Investing One's Money Wisely and Making It Grow, Treatise on the Theory and Practice of Commercial Papers and the Operations of the Stock Exchange, General Theory of the Operations of the Bourse, Principles of Commercial Sciences, Accounting: Theory, Practice and Education,* and even a publication on how to bet on horses. His association with the revered James de Rothschild, which actually may have been no more than a short-term interlude, served him well in his publishing endeavors, and he proudly displayed this alliance as a mark of quality on the frontispiece of his printed works.

In 1873 Lefèvre tried to obtain a teaching position at École des Ponts et Chaussées, then—as now—one of France's leading engineering schools. The school's library still has two eighty-page brochures by Lefèvre on its shelves, initially appended to his application, in which he proposed a syllabus for a course on political economy. Unsuccessful at Ponts et Chaussées, he persisted nevertheless in his efforts to have the theory of finance and commerce taught at universities and managed to introduce a course on advanced financial techniques at the Institut Polytéchnique (not to be confused with the much more famous engineering school École Polytéchnique). His last published work, *Theory, Practice, and Education of Commerce,* dates from 1885. The place and date of Lefèvre's death are unknown, but since nothing further was heard of him after that date, he must have died not long thereafter.

When Lefèvre began publishing scholarly books and articles in the early 1870s, the state of financial theory was less than satisfactory. "Theses . . . which promise much more than they actually deliver, provide no help at all in understanding the intricate mechanism of the operations that they claim to explain," he wrote in the preface to one of his early works. Inspired by biology, mechanics, and geometry, he aimed to discover the scientific laws that govern the circulation of commodities, capital, and securities. To this end, he made use of two insights. The first, stemming from his studies of the natural sciences, was to compare the economy and the flows of capital and goods with the human body and its organs. The second was an ingenious way of visualizing the position in which a speculator finds himself after one or several trades on the stock market.

In Lefèvre's model of the economy, the stock exchange is like the heart, which keeps blood moving through the veins. Two "organs" then influence the functioning of the flow: the government and speculation. According to Lefèvre, the government may be compared to the brain, which thinks and combines and regulates the flows but also hesitates, tires, and occasionally falls asleep. Speculation, in contrast, is like the body's nervous system. It provides the momentum that keeps commodities and capital in continuous motion.

Similar to human organs that keep temperature, pulse, blood pressure, and other vital variables stable throughout the body, markets ensure that prices of commodities and securities are constant across distances. The economy does this through arbitrage, an operation that exploits small differences in price, thereby making them disappear. Arbitrageurs, sensing an opportunity to make a profit, purchase securities wherever the price is low while simultaneously selling them wherever the price is high. With varying demand and supply, prices adapt in the different places, and differentials soon disappear. In early times, it was difficult to exploit arbitrage opportunities. The Rothschild family was able to profit from price differentials only because they owned a fast and efficient communications network, composed mainly of couriers on horseback. But with improved methods of communication—an electrical telegraph was developed and patented in the United States by Samuel F. B. Morse in 1837, Thomas Alva Edison devised a two-way telegraph in the early 1870s—any broker in Paris could be informed without delay if the price of a security in, say, London was lower or higher than at home, and would dispose accordingly. Thus financial markets ensure that the same price reigns everywhere.[2]

Given the activities of astute arbitrageurs, the principal problem of commerce and trade is not the variability of prices across space, Lefèvre maintains, but across time. As time passes, uncertainty enters. Much can happen in the period between the purchase of raw materials, the production of goods, and their eventual sale. Tastes may change, production costs fluctuate, prices vary. These events, the magnitude and even direction of which are uncertain, introduce randomness into every commercial activity and entail considerable risks for the producer. The same holds for investors, of course: it is the essence of speculation that the prices of securities vary between the time of their purchase and the moment of their liquidation. Lefèvre concluded that the bourse and speculation, the economy's heart and nerves, are the "social organs" that deal with the uncertainties arising from the passage of time. In particular, the *marchés à terme,* futures markets, allow producers to avoid the risks due to the uncertainties of the future. Farmers, for example, can insure themselves against the vagaries of time by selling their produce in a futures market at a predetermined price.[3] By agreeing in advance on a price, a quantity, and the date of delivery, they shift the risk to wholesalers or financiers who are willing and better able to assume the perils of changing prices.

Futures markets also exist for securities. Thus a speculator can agree with a willing partner on a certain price in the future, irrespective of what should happen to the value of the stock in the interim. The possibility to assure oneself of a definite price, thus becoming immune to the risk of excessive decreases or increases, provides a stimulus to economic and financial activities.

There is another kind of market and it is actually the theme of this book: the *marché à prime,* or options market. Though similar to futures markets, options markets exhibit a slight but significant difference. A futures trade is a commitment by both parties to fulfill their obligations. When the agreed-upon date arrives, the seller must deliver the securities (or commodities), at the agreed-upon price and the buyer must take possession. An option, on the other hand, gives the buyer a choice. He can insist on delivery of the security, if he so desires. But if circumstances make it advantageous for him not to do so, he is allowed to opt out of the deal. For the right to either insist on delivery or to refuse it, he pays the seller of the option a fee, called a premium (*prime* in French), which is the price at which the option is bought.

As we will see, the option can be thought of as insurance against too much volatility in the underlying security, and the money that one pays for the option can be considered the insurance premium. But there is a problem. What is the correct price of an option? How much should a buyer be willing to pay for insurance against volatility risk? There are different ways to assess

a security's value, for example, by the assets that make up the firm's property, or by its earnings potential. But the value of an option is unclear. Demand and supply somehow result in a market price, but does that price
reflect the option's true value? This is the main question that concerns us
in this book.

In three articles titled "Physiologie et mécanique sociales" (Physiology
and social mechanics) published in 1873 and 1874 in the *Journal des Actuaires Français,* Lefèvre analyzes this new financial instrument called *opération à prime,* or "options trade" in modern English parlance. He illustrates
the intricacies of options trades with the predicament of a *boulanger,* a
baker, who needs to assure himself of a constant supply of ingredients for
his croissants and baguettes. The price of flour on the market can fluctuate,
which entails great uncertainty for the poor baker. If it goes down, his profits will increase and he will be happy. But if the price rises, he may incur
losses and even go bankrupt. Ideally, the *boulanger* would like to have the
best of both worlds, profiting from low prices but keeping a lid on the maximum price. Surprisingly, this is possible. In exchange for a small fee, an insurance premium so to speak, there are speculators who are willing to
provide him with a safety net while letting him profit from low flour prices.

Let's say that on February 1 the *boulanger* purchases an option to buy
flour that is to be delivered on April 1, the exercise date, at the exercise
price of 5 francs a kilo. For the right to demand delivery of the flour he pays
a fee of, say, 20 centimes to a speculator. If, on April 1, the price turns out
to be lower than 5 francs, the baker lets the option expire without making
use of it since he can get the flour more cheaply on the open market. The
speculator pocketed the initial premium of 20 centimes without having to
do anything in return. But in case the price is, say, 5 francs 50 on April 1,
the baker will insist on delivery of the flour and only pay the agreed-upon
price of 5 francs. Not having warehouses or stock, the speculator must purchase the flour on the open market. He buys it for 5 francs 50 and delivers
it to the baker, who pays him the agreed-upon 5 francs. Taking into account
the 20 centimes that he received on February 1, the speculator thus made
an overall loss of 30 centimes. The baker, on the other hand, assured him-

self of a maximum price of 5 francs in exchange for the 20 centimes premium. We can see why the baker would want to enter such a transaction, but who would be willing to provide him such an option? A speculator, of course, who believes that the flour price will be lower than 5 francs on April 1. If he is right, and the price does end up below the exercise price on April 1, he will pocket the premium as an easy profit.

So far, there is nothing new in Lefèvre's exposition. The existence of futures and options markets, in spite of their dingy reputation, attests to the fact that they fulfilled a pressing need. Illicit but tolerated, their essential role in keeping operations on the bourse running smoothly and efficiently was known to all practitioners and observers. Regnault had already written about futures markets a dozen years previously. It was Lefèvre's next insight that would provide his claim to fame. It equipped speculators with a tool that allowed them to assess their financial position quickly and accurately. Even today, with computers and financial calculators abounding, Lefèvre's method is still the standard way of looking at, describing, and understanding the behavior of stocks and bonds. Twenty-first-century students are introduced to Lefèvre's graphs in their first classes on financial markets.

As I pointed out above, until about the middle of the nineteenth century, the theory of economics consisted mostly of descriptive texts, which stifled theoretical progress. But this was about to change. Under the influence of accountants, whose profession made them unafraid of using a quantitative approach, economics became more mathematical. Lefèvre was one of the pioneers of this revolution.

Lefèvre's geometrical approach employed no equations or formulas. But even though the mathematical component is hidden behind graphs, the approach clearly has mathematical underpinnings. He justifies his soft approach by appealing to geometric intuition: "While the practice of arbitrage is an interesting application of ordinary algebra," he wrote in 1879, "the practice of speculation is obliged to borrow from geometry in order to explain the combinations that are impossible to understand clearly with the help of arithmetic, algebra or plain speech alone." After all, if a picture is worth a thousand words, a graph is worth a few dozen mathematical symbols.

Lefèvre aimed to make trading on the stock exchange more scientific. In the breathtaking pace of the bourse, traders had to be as efficient as possible. With brokers at the ready around the *corbeille,* the enclosed area in which trading took place, gesticulating widely while shouting buy and sell orders, there was no time for speculators to ponder investment strategies. They operated according to gut feelings when deciding which risks to take. (Or they did not, in which case they would lose piles of money.) As long as few securities were traded (only 3 in 1800), keeping an overview of one's investments was relatively simply. But with the number of stocks and bonds quoted on the bourse constantly increasing (nearly 700 in 1879), the situation became unmanageable. Keeping tabs on one's exposure to risk became increasingly difficult.

Lefèvre sought a tool that would permit traders to analyze their position quickly and respond instantaneously to changing circumstances. However complex the holdings in the cash and the futures markets, and whatever prices are quoted, the rules should be so simple that any man on the street could easily apply them.

Explaining his ingenious method requires a bit of geometry and some graph paper. We will depict a coordinate system where the x-axis—drawn from left to right at half height on the graph paper—represents the various prices a security can assume. The y-axis represents the speculator's profit or loss after he sells the security.

Assume that a speculator buys a security for 100 francs. (In the following discussion, I sometimes omit the denomination francs.) If the price on the market is 100 when he eventually sells the security, he comes out even: his profit is zero. But for each franc above 100, the speculator's profit increases by 1 franc, for each franc below the acquisition price, his loss increases by 1 franc. Thus if he sells the security for 101, he makes a profit of 1 franc, at 102 he makes 2 francs. If he sells for 99 he loses 1 franc, if he sells for 98 he loses 2 francs and so on.

Nothing really surprising here, and Lefèvre's method is easily illustrated on our graph paper. Draw a line, inclined by 45 degrees, which crosses the x-axis at the purchase price (i.e., at 100). (See Figure 1.) From the graph

FIGURE 1: SELLING A SECURITY

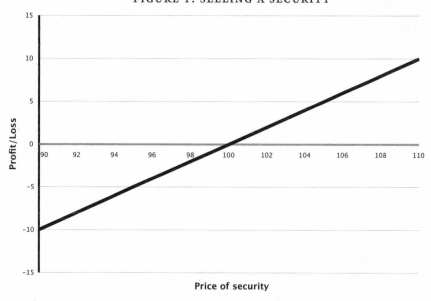

you can easily read off the profit or loss, as it depends on the sales price. For example, at a sales price of 105, the line shows a profit of 5 francs; at a price of 93, the line lies in the negative part of the graph and indicates a loss of 7 francs.

Now on to more complicated and interesting matters: the options markets. We return to our *boulanger* who concluded the options deal with the speculator for the delivery of flour at 5 francs. Let's first look at the situation from the baker's perspective. The price of flour on the market is displayed along the *x*-axis, the price that the baker pays along the *y*-axis. On February 1 the baker pays the speculator the premium of 20 centimes. On April 1, he buys the flour at the market price whenever the latter is below five francs. If the market price is above 5 francs, he uses his option and demands delivery from the speculator at a price of 5 francs. Thus if the market price is, say, 4 francs 60, the baker pays a total of 4 francs 80 (4.60 for the flour, 20 centimes for the option). If the price is 5 francs 60, he pays only 5 francs 20 (5 for the flour at the exercise price, 20 centimes for the premium).

On graph paper the situation is depicted by the kinked line in Figure 2. Below a price of 5 francs, the *boulanger* pays the market price plus the

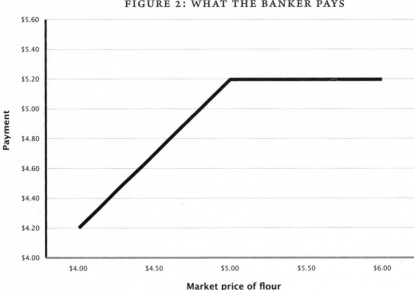

premium, and this is displayed by the left part of the line, which rises at an angle of 45 degrees. Above the market price of 5 francs, the baker incurs a constant cost of 5 francs plus the premium of 20 centimes. This is displayed as the horizontal part of the line.

What does the speculator receive? As long as the market price is below 5 francs, the baker does not use his option and the speculator gets to keep the premium of 20 centimes. This is depicted by the horizontal part of the kinked line in Figure 3. Things change when the price is higher than 5 francs. Now the speculator has to buy the flour on the open market at whatever price is demanded and deliver it to the baker in exchange for just 5 francs. He thus receives 5 francs, plus the premium of 20 centimes, minus what he paid for the flour on the market. He breaks even at a price of 5 francs 20. For each centime that the price increases, the speculator loses an additional centime. This is displayed by the right part of the line, which drops at 45 degrees.

Lefèvre's geometrical method works not only for commodities but also for options on securities. There are two kinds: options to buy a security, which today are called "call options," and options to sell a security, which

FIGURE 3: WHAT THE SPECULATOR RECEIVES

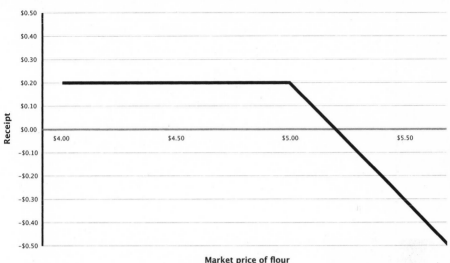

are called "put options." As in every trade, there must be two parties to the operation, and so we have four kinds of operators: buyers and sellers of call options, and buyers and sellers of put options. The profit and loss possibilities of each of these speculators can be depicted by a different kind of kinked line on graph paper. (In the appendix at the end of this chapter I give examples for each of these kinked lines and explain their significance.)

The advantage of the geometrical approach was that it provided an intuitive understanding of the inner workings of options trades. Indeed, Lefèvre's primary purpose in developing his graphical method was educational. But he and his colleagues quickly recognized that it had a very practical application. Speculators could instantly know the results they would incur, depending on the security's price, by reading their positions from the kinked lines on the graph paper.

With a single option, Lefèvre's method is not all that useful since it is easy to ascertain when one is about to make or lose money, even without checking whether the kinked line is in the positive or negative realm. The important advantage of the graphical method is that purchases and sales of multiple calls and puts can be combined into an overall, multiply kinked line. (See the appendix.) The combined position can then be graphed and a speculator holding an ensemble—however complex—of options on a security can ascertain with one glance the profits or losses that would result for any price that may be quoted for the security. If the price varies, he can react immediately and adjust his position.

Lefèvre's method proved to be very popular. The Syndicat des Agents de Bourse, the Paris brokers' organization, appointed a committee to study his graphs and was very complimentary. "Monsieur," they wrote in a letter to Lefèvre on February 24, 1874, "the syndicate listened with interest to a report of the members who had been charged to study your method of the bourse's operations. Since your graphs would be very useful to the members of the Compagnie [the brokers' association], it has been decided to purchase 60 copies [of the book *Principe de la science de la bourse*]. I ask you to deliver and charge them to our communal bursary." Since the Compagnie had sixty members, the syndicate's order ensured that each one of the brokerages would have a copy. Lefèvre proudly made the content of the letter known to the public and also had a large poster printed, on which his graphs were displayed. It could be obtained at the office of the Palais de la Bourse. In addition, he devised a mechanical contraption, an abacus, that helped do the financial calculations.

Regrettably, the very success of Lefèvre's method called another person onto the scene. The engineer Léon Pochet, a graduate of the École Polytéchnique and the École des Ponts et Chaussées, preoccupied himself mainly with hydraulics and steam engines throughout his career. Appointed chief inspector of agricultural irrigation by the French government, he is today mostly remembered, if at all, for a book on industrial mechanics. But at one point he was interested in the stock market and in 1873 he published the article "Géométrie des jeux de la Bourse" (Geometry of gambles on the stock market) in the *Journal des Actuaires*. For his explanations, he utilized the graphs that Lefèvre had invented without, however, mentioning them. Had he plagiarized Lefèvre?

In the paper, Pochet claimed that stock exchanges and forward markets are nothing but games of chance. Only two kinds of players stood a chance at turning a profit, he maintained, echoing remarks Regnault had made ten years earlier: professional players who stay abreast of political developments and other events that may influence stock prices, and big players who can directly influence the market by placing large purchase orders or by throwing vast amounts of stock onto the market. He then expounded on his geometrical method, which, lo and behold, was identical to Lefèvre's.

Pochet's work appeared in 1873, in the same volume of the same journal, but sixty pages before the first of Lefèvre's three papers. Thus the uninitiated might think that Lefèvre had plagiarized Pochet! Lefèvre was enraged. Citing two books and an article, and referring to himself in the first person plural, he points out that Pochet elucidated the principles of investments "with a method that we described and published in 1870 in various works and essays." Maybe Pochet really did discover the graphs independently or had been "inspired" by rumors he had heard about Lefèvre's work. Common courtesy would have demanded at least an acknowledgment by Pochet.

———

Lefèvre titled his essay "Physiologie et mécanique sociales" (Physiology and social mechanics) because he explained the workings of the stock exchange by analogy to the heart. Human societies are not just collections of juxtaposed individuals, he writes, or simply mechanisms. They are true organisms that can be found around the world in various stages of perfection. The heart of these organisms is the bourse. Thus the stock market occupies a more important place in the modern world than a simple gambling den, he reminds the reader, taking a quick swipe at Pochet's simplistic view of things. The bourse is the organ that keeps products, money, and stocks circulating. Science needs to understand the intricacies of this phenomenon.

In the course of 104 pages, spread over the three journal articles, Lefèvre then develops the graphical method. He ends the last paper in the series with an apology for the times when the bourse would fail to perform as expected. The human heart's function is to keep blood flowing, he writes. Whether the blood is good or bad, pure or contaminated, is none of the heart's business. It does not purify or filter; it just keeps on pumping. Similarly, the bourse should not be faulted if a market malfunctions now and then.

While Lefèvre is not exactly a household name nowadays, at least he was awarded a gold medal from the Société d'Encouragement pour l'Industrie Nationale (Association for the Encouragement of the National Industry) for his contributions to understanding the intricacies of the stock market. And Léon Say, four-time finance minister of France in the nineteenth century, compared Lefèvre's work to the discovery of descriptive geometry by Gaspard Monge.

————————

A SHORT PRIMER ON OPTIONS

There are two kinds of options: options to buy something—call options—and options to sell something—put options. Each option has two parties, a buyer and a seller. The latter is called the writer of the option.

We start with put options. Say the investor, Monsieur Ledoux, enters into a contract with his colleague, Monsieur Tavernier. Ledoux, the buyer, pays Tavernier, the writer, 5 francs, the premium, for the right to sell him one share of Immobilier et Céréales at a price of 100 francs on October 1. At the time when the contract is made, neither Ledoux nor Tavernier owns any shares.

Come October 1, Immobilier et Céréales trades on the bourse at 93 francs. A smiling Ledoux decides to exercise his option. He purchases a share at the market price of 93 and delivers it to Tavernier who—per contract—has to pay him 100 francs for it. Ledoux just made 7 francs profit on the trade, but since he paid Tavernier 5 francs at the outset, his overall profit is reduced to 2 francs. Poor Tavernier! After paying 100 francs to Ledoux, he can do nothing else to avoid any further uncertainty than to get rid of the unwanted share at the market price of 93 francs. He made a loss of 7 francs but can take

FIGURE 4: A PUT OPTION

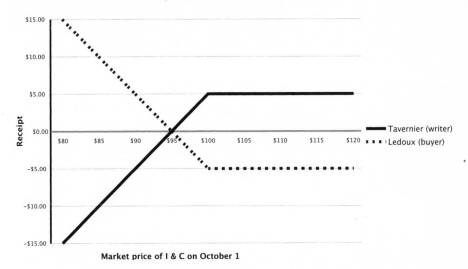

Market price of I & C on October 1

some comfort in the fact that he had received 5 francs when he sold the option. Thus his overall loss is 2 francs. The lower the share is traded on the exercise date, the higher are both Ledoux's profit and Tavernier's loss.

Now, if the company's shares are traded at, say, 101 francs on October 1, Ledoux certainly won't buy a share in order to sell it to Tavernier at 100. (Remember, an options contract gives the writer the right, but does not oblige him, to consummate the trade.) So Ledoux simply lets the option expire, writing off as a loss the 5 francs that he paid at the outset. Tavernier, on the other hand, breathes a sigh of relief; he just earned 5 francs without any further obligation. And nothing changes if Immobilier et Céréales is traded at 102, 103, or any other price higher than 100; Ledoux will always let the option expire and Tavernier always gets to keep the premium.

Let's use Lefèvre's graphical analysis. In Figure 4 the price of the share on October 1 is displayed along the *x*-axis, Ledoux and Tavernier's profit and loss potentials along the *y*-axis. For each market price of Immobilier et Céréales, the investor's payouts or losses can be read off the graph.

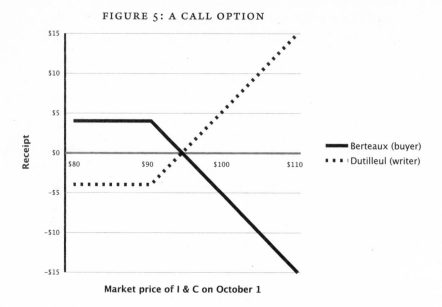

FIGURE 5: A CALL OPTION

Market price of I & C on October 1

On to call options, the options to buy a security. Again we may take Immobilier et Céréales as the security underlying the option. Monsieur Berteaux, the buyer of the call option, pays Monsieur Dutilleul, the writer, four francs in exchange for the right to buy a share of Immobilier et Céréales on October 1 for ninety francs. Come October 1, Immobilier et Céréales trades on the bourse at ninety-five. Rubbing his hands, Berteaux rushes to Dutilleul, pays him ninety francs in order to get one share. Dutilleul, who does not own any, must purchase it on the market for ninety-five and hands it over to Berteaux. He just lost five francs. But since he received four francs when the option deal was made, his overall loss is reduced to one franc. Berteaux takes the share for which he paid Dutilleul ninety francs, turns around and sells it on the market for ninety-five, making five francs on the trade. This payout is reduced by four, the initial premium paid to Dutilleul, to yield an overall profit of one franc. The higher the share price on October 1, the higher are both Berteaux's profit and Dutilleul's loss.

On the other hand, if Immobilier et Céréales trades at eighty-eight on the exercise date, Berteaux will certainly not rush to demand delivery of a share in exchange for 90 francs, since he could obtain it on the bourse for 88. He will simply let the call option expire. Dutilleul pockets the 4 franc premium that Berteaux writes off as a loss. Lefèvre's graphical analysis displays the situation in Figure 5.

As I said above, the ingenuity of Lefèvre's method is not only to illustrate the effects of a certain kind of option, but to make clear what may happen when multiple options are combined. One simply draws the kinked lines for each of the options. To arrive at the profits or losses of the combined portfolio, one then adds the results vertically at each price of Immobilier et Céréales. Let us say Monsieur Couret-Pléville enters the following five transactions, all for the same exercise date:

1) He writes a put option at the exercise price of 96 and gets a premium of 4.

2) He writes a call option at the exercise price of 102 and gets a premium of 2.

3) He buys a put option at the exercise price of 104 and pays a premium of 2.

4) He buys a call option at the exercise price of 108 and pays a premium of 4.

5) He buys another call option at the exercise price of 107 and pays a premium of 1.

Just looking at the data, nobody could even guess under which circumstances Monsieur Couret-Pléville would be making money from his portfolio and under which circumstances he would lose. Lefèvre's method makes it possible.

In Figure 6, the kinked lines, printed in gray, correspond to the above options. The financial effect of the portfolio—which consists of the combination of the five options—is shown as the multiply-kinked, bold line.[4] It is obtained by vertically summing the profits and losses of the five individual options at each possible market price

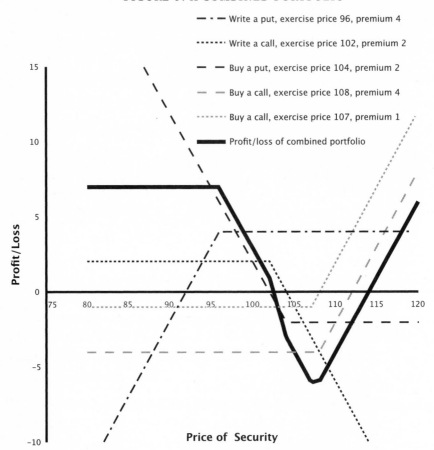

FIGURE 6: A COMBINED PORTFOLIO

— · — Write a put, exercise price 96, premium 4

······· Write a call, exercise price 102, premium 2

— — Buy a put, exercise price 104, premium 2

— — Buy a call, exercise price 108, premium 4

········ Buy a call, exercise price 107, premium 1

▬▬ Profit/loss of combined portfolio

of Immobilier et Céréales. It can now easily be seen when Monsieur Couret-Pléville will profit from his portfolio and when he will lose. He will always be making money, except if the market price of Immobilier et Céréales on the exercise date is between 103 and 114 francs.

FIVE

The Spurned Professor

THE STUDY OF FINANCIAL MARKETS BEGAN IN EARNEST at the turn of the twentieth century. The first underpinnings of a mathematical theory were developed in the doctoral thesis of a thirty-year-old French student of mathematics by the name of Louis Bachelier. In spite of receiving high praise from the committee of examiners, it was forgotten for about fifty years. Only when it was dug up by an American professor—who would later win one of the first Nobel Prizes for economics—did the world become aware of this path-breaking work.

Louis Bachelier was forced by circumstances to become a businessman in early life, but his love for mathematics and research eventually overcame all obstacles. He was born in 1870 as the eldest of three siblings into a well-to-do family in the northern French town of Le Havre. His father was a wine merchant, and his mother hailed from a family of bankers. When Louis was nineteen years old, tragedy struck the family. Both his parents died, his mother four months after his father. Louis, who had just passed his high school exams, was forced to care for his family and the family business. Thus the talented Bachelier was not able to pursue the two-year preparatory course for the *concours,* the highly competitive mathematics exams that are still de rigueur in France for budding scientists. Consequently he was barred from entering the elite universities, the École Normale Supérieure and the École Polytechnique. This shortcoming would haunt Bachelier all his life. Moreover, he was drafted into the French army for his compulsory one-year military service when he was twenty-one.

After his army service, Bachelier headed for Paris, where he found a job at the bourse in 1892. It is not known at which firm he was employed nor in what capacity. Each *agent de change* required an army of assistants if only to transmit messages, because even though the telephone had been invented some fifteen years earlier, it was not yet widely used. And there was much else to do. Was Bachelier one of the clerks rushing from the office to the *corbeille* with clients' orders, a bookkeeper entering numbers into ledgers, a registrar doing the paperwork? Whatever his position, it gave him the necessary experience with the inner workings of the market for financial securities. As we shall see below, he would put his knowledge to good use.

At the bourse, he got an excellent understanding of the cash market and the forward market. But Bachelier did not see himself as a stockbroker. As soon as he managed to put aside some money, he returned to the studies that had been interrupted after high school. Taking courses at the Sorbonne under Henri Poincaré, Paul Appell, Èmile Picard, and Joseph Boussinesq— some of the foremost mathematicians and physicists of their time—Bachelier received a solid grounding in pure mathematics and mathematical physics. In particular, he studied the theory of heat. The diffusion equation, which describes the circulation of heat, would later become one of the instruments that Bachelier used to analyze price movements on the stock market.

He was a mediocre student. In order to receive the first degree, comparable to something in between a bachelor's and a master's degree, he had to take exams in differential and integral calculus, mechanics, and astronomy. Bachelier got by, but only by the skin of his teeth, sometimes passing the exams after several attempts and often finishing bottom or next to bot-

tom of his class. Actually, this assessment may be too harsh, because graduating from the French university system was hard, and even coming in last
place was a respectable achievement. Moreover, Bachelier was at a distinct
disadvantage compared to his fellow students from the start. As already
mentioned, he had not had the opportunity after high school to pursue the
two-year preparatory course in mathematics at a *lycée*.

Eventually Bachelier received a degree in mathematics. The difficulties
along the way did not stop him from reaching for even higher academic
credentials. He went on to write a Ph.D. dissertation titled "Théorie de la
spéculation" (Theory of speculation). It dealt with . . . well, speculation.
And it was unlike anything the mathematicians at the Sorbonne had ever
come across.

Two and a half weeks after his thirtieth birthday, on March 29, 1900,
Bachelier had to defend his doctoral work. The thesis committee was composed of Poincaré, Appell, and Boussinesq. He acquitted himself well. As
was the Sorbonne's requirement at the time, Bachelier also had to defend
a second thesis, this one on fluid dynamics. Concerning the latter, Appell
reported that the candidate had a good command of the theory of movement of spheres in a fluid. But it was the thesis on speculation that elicited
the interest of the jury's senior member.

The evaluation, written by Poincaré and signed by all three men, was
nothing if not complimentary. It started with a caveat, however. "Le sujet
choisi par M. Bachelier s'éloigne un peu de ceux qui sont habituellement
traités par nos candidats." (The subject chosen by Mr. Bachelier is a bit removed from those traditionally treated by our candidates.) Its objective is,
Poincaré continued, the application of probability calculus to operations
of the bourse. One could imagine combinations of securities that would
guarantee a certain profit, but the laws of the market rule out such situa-
tions, he wrote. And if they do occur, they cannot persist. Furthermore,
while buyers of a security believe in its probable rise, sellers believe in its
fall. Hence the market overall expects price changes to be zero. And from
this, the famous law of errors of C. F. Gauss follows—the bell curve—provided that the price deviations are not too large.[1]

Poincaré termed Bachelier's derivation of Gauss's law, and in particular the comparison of stock price movements with the flow of heat, *fort original* (very original). He was quite surprised that the results of heat flow analysis were applicable nearly unchanged to the stock market, a problem so different from the one for which it was originally created. He expressed regret that Bachelier did not develop that part of his thesis further. Then he turned to the part that today we consider the central element of Bachelier's work, "la relation entre la valeur d'une prime et l'écart avec le ferme" (the relationship between the value of an option and the spread between the cash price and the price of the future's contract). This relationship can be empirically verified through observations at the bourse, Poincaré remarked, but warned that one should not expect an exact relationship. Whenever the relationship is violated, traders would play the market in such a manner as to reestablish the equilibrium.

Bachelier's thesis then moves on to a very difficult problem. "What is the probability that a certain stock price will be reached before a certain date?" At first Poincaré wanted to write that the problem "seems unsolvable" but then crossed out "unsolvable" and wrote that it "seems to give rise at first blush to very complicated computations." But the author manages to solve it, Poincaré stressed, through "short, simple, and elegant reasoning." The report ended with a curteous conclusion. "En resumé, nous sommes d'avis qu'il y a lieu d'autoriser M. Bachelier à faire imprimer sa thèse et à la soutenir." (In summary, we are of the opinion that Mr. Bachelier should be authorized to have his thesis published and endorsed.)

But then came the big disappointment. Doctoral theses in France were given evaluations of "not worthy of consideration," which means exactly that; "passable," which means in English what it says in French; "honorable," which means "somewhat better than okay"; and—the real distinction—*très honorable*. Bachelier only received the second highest rating. Appell summarized the committee's verdict: "La faculté lui a conféré le grade de Docteur avec mention honorable." (The faculty bestowed on him the degree of doctor with an honorable mention.) The less than outstanding assessment would haunt Bachelier for the rest of his life. But with a thesis topic outside pure mathematics, the jury apparently felt it could not go any higher than that. Even the publication of the thesis as a sixty-six-page article in the prestigious *Annales scientifiques de l'École Normale Supérieure* was of little consolation to Bachelier.

At the beginning of the twentieth century, it was very difficult to get a teaching position at a university in France even under the best of circumstances. In the whole country there were only some fifty posts available for mathematicians. Without the grade of *très honorable* on one's doctoral thesis, a job search was practically hopeless, especially if one did not belong to the old boy networks of *Normaliens* and *Polytechniciens*.[2]

Even though Bachelier could not find a teaching job, he never gave up. Supported by grants and fellowships, he continued his scientific work, developing the theory of stochastic differential equations (more on that in later chapters) using the language and concepts of gambling. Starting in 1909, Bachelier gave lectures at the Sorbonne . . . for free. Particularly noteworthy was the course Probability Calculus with Applications to Financial Operations and Analogies with Certain Questions from Physics. In 1912 he published "Calcul des probabilités," which according to many contemporaries surpassed Pierre-Simon Laplace's famous treatise on the subject of exactly a century earlier, and in 1913 the Sorbonne finally started paying him a salary. A year later, his book *Le jeu, la chance et le hasard* (Gambling, chance, and risk) appeared and went on to sell more than seven thousand copies. Also in 1914, the Council of the University of Paris finally considered offering Bachelier a permanent appointment.

Just as things were starting to look up, bad luck struck again. World War I broke out, and the forty-four-year-old self-employed scientist was drafted into the army. That spelled the end of his hopes for a post at the Sorbonne. Bachelier remained on active duty throughout the war. Upon demobilization as a lieutenant on December 31, 1918, he assumed replacement positions at the universities in Besançon, Dijon, and Rennes, hoping that a position suitable to his talents and interests would eventually turn up. In the meantime he never ceased researching and publishing.

His experience in Dijon was especially galling. The university considered turning his temporary appointment into a permanent position and requested an evaluation of Bachelier's scientific work from Maurice Gevrey, who held the mechanics chair at the university. Gevrey, an alumnus of Normale Sup and not keen on getting an upstart like Bachelier as a colleague, inspected Bachelier's work superficially and thought that he spotted an error in one of the papers. Since he was not an expert in probability, he requested an opinion of Paul Lévy, a colleague from their student days and now a well-known professor at École Polytechnique. Lévy looked into the matter without giving it the attention it deserved, confirmed the error, and Bachelier was passed over for the position. The lucky candidate who did get the job was a brilliant mathematician by the name of Georges Cerf who—you guessed it—had been a student at Normale Sup.

Bachelier was furious, especially since he never got a chance to defend himself against the accusations. He could have easily invalidated them, based as they were on a misunderstanding. In the paper reviewed by Gevrey and Lévy, he had taken a notational shortcut that he felt no need to justify, although he had done so in his doctoral thesis. That was not to the liking of rigorous mathematicians, but quite in the spirit of mathematical physics and hence fairly acceptable. In an open letter to colleagues, Bachelier deplored the fact that his career had been blocked in an extremely unfair way by Lévy, who had not even bothered to familiarize himself with his work.

Years later, Lévy realized that he had wronged Bachelier. While reading a fundamental paper by the Russian mathematician Andrei Nikolaevich Kolmogorov, whom he greatly respected, he became aware of his error. In a paper in 1931, Kolmogorov had singled out Bachelier's contributions for special praise . . . and for criticism. "As far as I know, Bachelier was the first to make a systematic study," he wrote, but then, unfortunately, went on to note "that Bachelier's constructions are by no means mathematically rigorous." However, even qualified praise by such a famous mathematician prompted Lévy to give Bachelier's paper a closer reading. He quickly realized his own error.

To Lévy's credit, he wrote a letter apologizing to Bachelier, and the two men reconciled shortly before Bachelier's death. In his memoirs, published in 1970, Lévy acknowledges Bachelier's pioneering work albeit with a qualification. "If the work of Bachelier, which appeared in 1900, has not attracted attention, it is because, on the one hand, not everything was interesting . . . and because, on the other hand, his definition was at first incorrect." He then explains why he ignored it at first. "I myself did not think it useful to continue reading his paper, astonished as I was by his initial mistake. It is Kolmogorov who quoted Bachelier in his 1931 article . . . and I recognized the injustice of my initial conclusion."[3]

After the debacle at the university in Dijon, Bachelier continued doing odd jobs teaching here and there and doing research. In a 1920 paper, for example, he investigates whether the first 707 decimal digits of the number π—that is how many were thought to be known then—can be considered to follow each other in random order. (His answer: yes.)[4] In 1927 a permanent position was proposed to the fifty-seven-year-old Bachelier by the University of Besançon. Bachelier accepted the offer and remained there until his retirement ten years later. During his tenure he did not publish anything, busy as he was preparing his lectures. After his retirement, he moved to Bretagne with one of his sisters and had three books on probability theory printed at his own expense. Bachelier published his final paper in 1941. He died on April 29, 1946.

While largely ignored by mainstream mathematicians, economists, and finance professionals during the first half of the twentieth century, Bachelier's work did get mentioned in France. In 1908, a book on the theory of probability and its applications by the Vicomte Robert de Montessus de Ballore, professor at the Faculté de Sciences in Lille, cites Bachelier and so does a book on the theory and practice of financial operations, in the same year, by the professor of finance at the University of Paris, André Barriol. Two years later, the speculator Maurice Gherardt presented a model based on Bachelier's theories in a "how to get rich quickly" manual.

Of course, Kolmogorov (more on him in Chapter 10), and eventually also Lévy, were familiar with his work. Another early but cautious adopter of his ideas was Lord John Maynard Keynes, the great English economist, who reviewed one of Bachelier's books. "On what he has accomplished it is not very easy to pass judgment. The author is evidently of much ability and perseverance, and of great mathematical ingenuity; and a good many of his results are undoubtedly novel. Yet, on the whole, I am inclined to doubt their value and, in some important cases, their validity." But then he qualifies his assessment: "I do not make this judgment with complete confidence, for the book shows qualities of no negligible order." What these qualities are will be seen shortly.

SIX

Botany, Physics, and Chemistry

B ACHELIER'S PH.D. DISSERTATION, AS WELL AS MUCH OF his later work, concerned price movements on the stock market, how the peaks and troughs, the ups and downs follow each other. He did not really invent a new concept. As it turned out, an English botanist, of all people, had seen it before . . . in a totally different context. And he had not been the first one either. He simply rediscovered a phenomenon that had been observed by many curious scientists previously.

The discovery that would immortalize the British army doctor and botanist Robert Brown's name was made by the British during the first half of the nineteenth century. Surprisingly, his claim to fame among the general scientific public was not the classification of a new plant, although the International Plant Names Index does list 10,753 species, from *Abelia chinensis* to *Zygochloa paradoxa,* that are associated with him. It was a discovery outside his immediate field of interest that made him a household name.

Born in Scotland in 1773, Brown studied medicine at the University of Edinburgh, and at age 23 joined the Fifeshire Regiment of Fencibles as assistant army surgeon. The regiment, a militia for the defense of the home front (hence the name, which derives from "defensible") was stationed in Ireland. It was Brown's duty to attend to the various bodily conditions and ills of the men. But his schedule was not very demanding. Since the 500 to 1,000 soldiers and officers were, for the most part, able-bodied men, it sufficed to see patients daily between the hours of 1:00 PM and 3:00 PM. This cushy schedule left Brown sufficient leisure to indulge in a hobby, collecting plants. Every day, from after breakfast until lunchtime, and again in the afternoon after his office hours, he gathered leaves and shrubs and worked on botanical essays. Most evenings, when not socializing or dining out, Brown continued his scientific work until late at night.

In 1798 Brown was in London to enlist new recruits for his regiment. He was introduced to an eminent botanist, Sir Joseph Banks, who took an interest in his botanical pastime. Two years later, the wealthy naturalist and science patron offered the army doctor the possibility to pursue his hobby full-time. The ship *Investigator* was to chart the coastline of Australia, and Brown was invited to join the naval expedition as resident botanist. He accepted and for the next four years he collected, studied, and classified about 3,400 plant species, more than half of which were unknown. Unfortunately the boat that was to bring the specimens back to England was shipwrecked on the Great Barrier Reef off the coast of Queensland in northeast Australia, and a large part of his collection was lost.

Brown returned to England in 1805 with a cornucopia of drawings, notes, and specimens. He spent the next five years describing his findings, eventually publishing them as *Prodromus Florae Novae Hollandiae et Insulae Van-Diemen* in 1810 (Introduction to the Flowers of New Holland and Van-Diemen's Land). New Holland was the name used for Australia, and Van-Diemen's Land is today's Tasmania. In 1806 Brown was named "clerk, librarian, and housekeeper" of the Linnean Society, whose president he was from 1849 to 1853. When Sir Joseph died in 1820, he left his home and library to Brown, specifying that he should be curator of his collections.

Brown then negotiated the integration of Sir Joseph's scientific legacy into the British Museum's botanical collection, whose keeper he became until he died on June 10, 1858.[1]

The reason that Brown's name is known today among nonbotanists—physicists, statisticians, economists, finance professionals—is an event that occurred sometime in 1827. Brown was inspecting the pollen of the species *Clarkia pulchella,* a pink flower of the evening primrose family. He was trying to understand the mechanism of fertilization in flowering plants. Having worked on the ovum, he now wanted to inspect the pollen, in particular the structure of the pollen grains. To do so, he suspended the powdery substance in water. Peering at the petri dish through his microscope, he noticed something strange. The particles were constantly in motion, performing "rapid oscillatory motion." Without ever stopping, they danced on the water's surface—like strung-out ravers on a disco floor, we might say today. A year later, he published his finding in *Philosophical* magazine under the title "A Brief Account of Microscopical Observations Made in the Months of June, July, and August, 1827, on the Particles Contained in the Pollen of Plants; and on the General Existence of Active Molecules in Organic and Inorganic Bodies."

At first, Brown thought that currents in the water or the fluid's gradual evaporation could be responsible for the nervous, jittery movement, or the gravitational attraction between the particles themselves. But he soon convinced himself that this was not the case. Even if a drop of water containing a single particle was immersed in oil, the movement continued unabated. So, were the pollen grains in motion because they were alive? Could the restless movement in the petri dish be a sign of the true origin of life? This is what his predecessors and contemporaries thought. Brown was not the first to observe the microscopic activity. Among the minuscule hullabaloo's earlier spectators he mentioned an international assortment of naturalists: the Dutchman Antony van Leewenhoek, the Frenchman Georges-Louis Leclerc Comte de Buffon, the Englishman John Turberville Needham, the Italian Lazzaro Spallanzani, the Irishman James Drummond, and the German Wilhelm Friedrich Freiherr von Gleichen-Russworm. All of them believed that the movements they saw under the microscope were associated with the mechanism of life.

Brown was very careful in his experiments. He was also very skeptical. To rule out the origin-of-life hypothesis, he performed an ingenious test. He carried out the same experiment again, but this time with dead matter. Taking some rock, he ground it into dust and sprinkled it onto water. Things just don't get any deader than that. Lo and behold, when the dead dust was observed under a microscope, it performed the same jitterbug dance as the pollen grains. So whatever it was that caused the dust's movement, intrinsic vital forces could be eliminated as a possible cause.

Actually, Brown could have resorted to an earlier experiment to rule out vital forces. More than forty years previously, in 1785, another botanist, the Dutchman Jan Ingenhousz, had observed powdered charcoal jiggling on the surface of alcohol. Since charcoal is just as dead as rock, it was obvious that the jiggling was not caused by life. Animate forces thus ruled out, a new theory was required to explain the movements, but neither Brown nor Ingenhousz provided it. As far as priority goes, Ingenhousz was the first to float dead matter on the surface of a fluid. Nevertheless, the phenomenon became known as "Brownian motion," and a good thing that was too, since "Ingenhouszian motion" would be quite a mouthful.

Brown had observed that the particles' motion was very irregular. Experiments conducted throughout the nineteenth century showed that two particles appeared to move independently and that the motion was stronger with smaller particles. Furthermore, it appeared that Brownian motion was more active the less viscous the fluid and the higher the temperature. Above all, experiments conducted over a period of twenty years showed that Brownian motion never ceases. But while more and more characteristics of Brownian motion were established, the reason for the movements remained unknown.

So what is it that makes particles jiggle? In order to not unduly tax the reader's patience, let me pull the rabbit out of the hat right now. The reason for the jiggly-jittery movements is that the molecules of the liquid on which the particles float slam into them from all sides, thus pushing them around.

The molecules come randomly from all directions. When more of them push from one side than from the other, the particle moves a minute distance sideways. The fact that higher temperatures lead to more vigorous motion also fits very well with this theory because the warmer the fluid, the faster the molecules.

The explanation seems very plausible. Actually, the basics of this idea had been around since about the fifth century BC, when the Greek thinker Democritus postulated that all matter consists of atoms. In 1808 schoolteacher John Dalton proposed that all matter is made up of a finite number of different atomic species. He also maintained that there are simple atoms and compound atoms, the latter being molecules.

At the turn of the twentieth century, the existence of atoms and molecules was not yet believed by many scientists, let alone established as a verified theory. The great physicists Ludwig Boltzmann and James Clerk Maxwell did postulate atoms in their models, but they used them only for didactic purposes, as convenient allegories. The virtual little balls were figments of their imagination, conceived only to illustrate their move from classical to statistical thermodynamics, but not to be taken literally.

Then Albert Einstein came along.[2] Among the revolutionary feats of this towering genius, one achievement is sometimes overlooked—exactly the one we are interested in here. Einstein did more than just create the theories of special and general relativity; he may also be rightly considered one of the fathers of the atomic theory of matter. It started with a *Gedankenexperiment,* the thought experiments that Einstein made famous. If we assume that atoms exist, Einstein thought to himself, what would happen to minuscule bodies suspended in a liquid?

Actually, he had partly covered this ground in his Ph.D. dissertation, "Eine neue Bestimmung der Moleküldimensionen" (A new determination of molecular dimensions). In this dissertation, which was initially rejected by the University of Zürich as being too short—just seventeen pages—but then accepted after Einstein added one more sentence, the young physicist investigated the size of the conjectured molecules. He did so purely on a theoretical level and then checked his reasoning, using available data on sugar dissolved in water. His estimate of the size of the imagined balls was about three-quarters of a millionth of a millimeter. The best estimate today is about one-tenth of a millionth of a millimeter, so in terms of orders of magnitude Einstein's reasoning was not far off.[3] And remember, when Einstein wrote his article, atoms and molecules were not even known to exist.

In 1905 Einstein published another paper on the physics of the undiscovered atoms in what is probably the world's most famous volume of a scientific journal, number 17 of the *Annalen der Physik*. It contained Einstein's paper "Über die von der molekularkinetischen Theorie der Wärme geforderte Bewegung von in ruhenden Flüssigkeiten suspendierten Teilchen" (On the motion of small particles that are suspended in liquids at rest, which is required by the molecular-kinetic theory of heat), nested in between his essay on special relativity (400 pages before) and his essay on the photoelectric effect (330 pages later). It is the middle article that interests us here.

"In this paper it will be shown that, according to the molecular-kinetic theory of heat, microscopically visible bodies suspended in liquids must perform movements—as a result of the molecular motion due to heat—that are of such magnitude, that these movements should be easily observable by microscope," Einstein wrote. He ended the first paragraph with a noteworthy remark. "It is possible that the motion discussed here is identical to the so-called Brownian molecular movement, but the information available to me is so sketchy that I cannot form an opinion about this." As it turned out, it was, in fact, Brownian motion that Einstein envisaged in his *Gedankenexperiment*. He had invented the phenomenon as a theoretical concept before he found out that it really existed.

Einstein saw his prediction as a test. Should the movements envisaged by him in fact be observed, he wrote, it would constitute a good argument for the conjecture that the speed of molecules is the source of heat (the molecular kinetic theory of heat). He did not say that the floating bodies move about erratically because they are bombarded by the fluid's molecules. After all, he had never seen, and only vaguely heard of, Brownian mo-

tion. But by applying the physical laws for osmotic pressure and for particles moving in a viscous medium, Einstein reasoned that they must be moving about.

In his *Gedankenexperiment,* he found something quite novel. He identified the irregular motion of the bodies in the fluid due to the diffusion process not as ordinary movements but as a random process: within each time
period, the bodies move a certain random distance. The distances are governed by the familiar bell-shaped curve, the Gaussian error distribution.
Thus he used the diffusion equation not as a description of the distribution
of a substance in a fluid but as the probability distribution of the bodies'
displacements.

The most significant consequence of Einstein's prediction was that, if
true, it gave a justification for the atomic theory of matter. For our purposes,
a different aspect of the paper is of interest. How far would the floating
body move, on average, during a given time period? Brown and others had
observed the jiggly, jerky movement, without knowing what it was and
without analyzing its speed. Mainly that was because they had no good definition of speed.

So Robert Brown observed and Einstein predicted (maybe "postdicted"
would be a better term, since Brown preceded Einstein) that bodies suspended in a liquid perform random movements in all directions. How far
would the bodies move? To answer that question, let us assume for sim-
plicity, as Einstein did, that the movement takes place only along one axis.
For the purpose of illustration, think of a drunkard lolling around a lamppost, and then tottering along the road. He sometimes takes a step forward,
sometimes backward. Lurching to and fro in this manner, will he ever get
home? And if so, how long will it take? Well, being random, the number of
forward steps is, on average, equal to the number of backward steps. Hence,
one may believe that the drunkard never leaves the neighborhood of the
lamppost. He should always return to the start of his tour, and hence his
average speed should be zero, right?

Wrong, said Einstein. He knew, of course, that the number of forward
and backward steps are equal, on average, and that the covered distance,

computed by adding and subtracting the steps, therefore sums to zero. But he considered a different measure of distance. To consider averages, one must, after all, observe not one but many drunkards. Stumbling along, some of them will be crowded around the lamppost, but many more will end up either at some distance to the left or at some distance to the right.

True—on average—they will all assemble at the lamppost and—on average—they will not have moved at all. But most individuals will have moved away somewhat. It is like saying that the average family has 2.3 children. Some families have two children, some have three, some have none, others have one or four, and so on. There is no family that actually has 2.3 kids. And yet the *average* family *would* have 2.3 kids.

So where does that leave us? If we do not care whether it's to the left or to the right that the drunkards move, if we are interested only in the distance from the lamppost, regardless of the direction, then a different manner of measuring average distances must be sought.

Einstein found it by realizing that the law of particles moving in a viscous fluid can be combined with the law of osmosis, which governs the movement of solvents (e.g., water) through semipermeable membranes into a more concentrated solution. This insight, coupled with some mathematical manipulations, led him to the appropriate measure of the distance that the particle covers. It is not simply the sum of the forward and backward steps. Einstein's key sentence was, "Die mittlere Verschiebung ist also proportional der Quadratwurzel aus der Zeit." (The mean displacement is thus proportional to the square root of time.) Wow! Is this not what Regnault had noticed, four decades earlier, concerning the movements of stock prices? But we are again getting ahead of ourselves.

Let's assume that the drunkards take a one-meter step every second, either to the right (+1) or to the left (−1). First, every step is squared so that both negative and positive steps give positive numbers; we now have a sequence of +1's. Then the elements of the sequence are summed and, to countermand the squaring, the square root is drawn. That's the solution. Take, for example, sixteen steps taken randomly in either direction. Squaring and summing: 16; drawing the square root: 4. So after sixteen seconds, the drunkards will have moved, on average, four meters from the lamppost. They may not get home immediately, but on average, given sufficient time, they will cover the required distance—possibly in the wrong direction.[4]

Einstein also gave some verifiable numerical predictions. With bodies of a diameter of one-thousandth of a millimeter, floating on water of 17 degrees Centigrade, the particle should move about a six-thousandth of a millimeter per minute.[5] He ends the paper with an appeal to colleagues: "It is to be hoped that a scientist will soon succeed in deciding the question raised here, which is so important for the theory of heat." The question was so important because verifying his predictions would decide whether the classical or the molecular-kinetic theory of heat was correct. And that, in turn, would decide whether matter was made of molecules.

Unbeknownst to Einstein, a physicist in Poland had already been on the same track for quite a while. Marian Smoluchowski, born in 1872 in a suburb of Vienna, was seven years older than Einstein.[6] His father, a lawyer and member of the chancellery of Emperor Franz Josef I, sent him to the Collegium Theresianum, a prestigious school in Vienna for the children of high-ranking officials, and the boy finished his high school diploma with distinction. He went on to study physics at the University of Vienna and after the obligatory stint in the army received his Ph.D. Remember Bachelier, whose doctoral thesis got the so-so commendation of honorable? Smoluchowski was quite at the other end of the merit ladder. His was a *promotio sub auspiciis imperatoris* (graduation under the auspices of the emperor). It can't get any better than that. The highest distinction of the educational system of the Habsburg Empire, it meant that the candidate had passed all exams since high school with the highest distinction. It also meant that the emperor would send a personal representative to the awards ceremony who would present the honored student with a diamond ring, engraved with the imperial initials. A university could hold at most one such ceremony a year. Smoluchowski was off to a very promising start, indeed.

He spent the next few years at famous laboratories in Europe—Paris with Gabriel Lippman (Nobel Prize 1908), Glasgow with William Thomson (later Lord Kelvin), Berlin with Emil Warburg (president of the German

Physical Association), and Vienna. In 1900 he was named associate professor of theoretical physics at Lwow University thus becoming, at age twenty-eight, the youngest professor in the Habsburg Empire. A year later he married Zofia Baraniecka. When he first met his future wife, he wrote, "I firmly retract what I had once written, namely that happiness is just the absence of unhappiness." Zofia would bear him a son and a daughter. In 1913 he moved to Jagellonian University in Cracow, becoming dean of the faculty of philosophy.

As an Austrian reserve officer, he was mobilized during World War I and put in command of an artillery detachment that had to guard a railway bridge between Vienna and Cracow. He was relieved of his duties after a few months. In June 1917, Smoluchowski was elected rector of Jagellonian University. Unfortunately, he would never assume his last post, succumbing prematurely to dysentery in September of the same year, at age forty-five.

Smoluchowski belongs to the select group of scientists who have a formula or a constant named after them. The "Smoluchowski equation" describes the evolution over time of the concentration of Brownian particles of a certain radius and mass that are suspended in a fluid with a certain viscosity and subject to an external force. Among the distinctions he received during his short lifetime were an honorary doctorate from the University of Glasgow in 1901, the Haitinger Prize of the Vienna Academy of Sciences in 1908, and—surprise—the Silver Edelweiss by the German and Austrian Alpine Society in 1916. Yes, Smoluchowski was not only an outstanding physicist but also an avid skier and able mountaineer. Together with his brother, he traced twenty-four new ascents in the Alps and climbed several difficult peaks, among them the legendary Matterhorn in Switzerland.

Smoluchowski was a romantic scientist; he loved art, music, and theater and played the piano. Modest to a fault, he never took advantage of his status as a professor. Albert Einstein would later write that "whoever was in a close relationship with Smoluchowski, valued not only his bright mind, but also a subtle and kind personality."

When Smoluchowski submitted his paper "Zur kinetischen Theorie der Brownschen Molekularbewegung und der Suspensionen" (On the kinetic theory of Brownian motion and of suspensions) to the *Annalen der Physik* in September 1906—four months after Einstein's article had appeared in the same journal—he had already been thinking about the phenomenon, and about the possibility that matter was made of molecules, for several years. He had not made his reflections public because he first wanted to verify them experimentally. Being unable to do so and prompted by Einstein's paper, he decided that it was time to publish his own findings, especially since he believed his method to be more direct, simpler, and more convincing than Einstein's. The paper was based on a talk that he had given to the Cracow Academy of Sciences two months previously.

Today it would take some nerve to suggest that one's arguments are superior in any way to Einstein's. But in 1906 the unknown patent clerk in Berne had just barely received his Ph.D., while Smoluchowski was already the thirty-four-year-old dean of the faculty of philosophy at the University of Lwow. Furthermore, he had just spent nine months at the Cavendish Laboratory in Cambridge with the famous physicist J. J. Thompson, who was about to be awarded the Nobel Prize.

The paper starts with an overview of experiments conducted during the previous eighty years. Particles had been suspended in liquids under varying circumstances, kept at boiling temperature for hours or in the dark for weeks. Never did the jiggly movements stop. Smoluchowski lists no less than twenty-four articles that mainly served to exclude various attempts to explain the phenomenon. Possible reasons like evaporation, convection, vibration, the intensity of illumination, the color of light, capillary energy, and electric energy were all rejected. The only possible cause for the "Brownian phenomenon" that seemed consistent with all available evidence was the movement of the conjectured molecules in the liquid.

He began by describing the observed motion. It was no oscillation. Rather, it was a sort of shivering—a *fourmillonnement*, or swarming about, as a French physicist said—which, in spite of the feverish agitation, resulted in only minute progression of the suspended bodies. Smoluchowski then took issue with the findings of Karl von Nägeli, a Swiss botanist whose main claim to fame was the discovery of the structures inside cells that would later become known as chromosomes. Nägeli had done some calculations about the momentum that a moving water molecule would impart on a larger body suspended in the liquid upon impact. He concluded that due to the impact of a molecule a suspended body would experience a velocity of only 0.00003 millimeters per second, which is far less than the speed that the jiggling Brownian particles display. Furthermore, since molecules come from all directions, even large numbers of collisions should not result in any movement, Nägeli maintained, because the impacts would balance each other out.

But the Swiss botanist had committed a grave error in his reasoning, Smoluchowski charged. It was the same error that a gambler would make if he betted on coin tosses and believed that he could never lose more than the stake of a single bet. As long as he throws the coin numerous times— so the faulty reasoning goes—the wins and losses would have to balance each other and the gambler would be out of the money at most with one bet. With the help of combinatorial methods, Smoluchowski shows that this is utterly wrong. After n coin tosses, n being large, the gambler would chalk up an average loss or gain of $0.8 \times \sqrt{n}$. To illustrate, after 100 tosses, the average gain or loss would be 8.[7]

It is the drunkard all over again. In the same manner that the drunks will not all end up at the lamppost but, on average, find themselves at some distance from it, so too gamblers will not necessarily balance their gains and losses at the end of their gambling sprees. The distance to the starting point is, on average, proportional to the square root of the number of steps (in the case of drunks) and tosses (in the case of coins).

Since the impacts from opposite directions do not cancel each other completely, what is the speed that is imparted to the bodies by the molecules? A body floating in water incurs about 10^{20} impacts per second. Hence, even though the vast majority of impacts cancel each other, there remains according to Smoluchowski's combinatorial argument a surplus of about $\sqrt{10^{20}}$, that is, 10^{10} impacts per second that point in the same direction. As per Nägeli's computations, each impact imparts to the body a speed of 0.00003 millimeters per second; hence the combined force of the impacts corresponds to 300,000 millimeters per second, which is about 1,000 kilometers per hour.

This is nearly as fast as the speed of sound and thus many orders of magnitude faster than what is observed under the microscope. Smoluchowski quickly explains the discrepancy. The faster the body already moves in one direction, the less likely is it that additional molecules will hit it in the same direction. He then deduces from the fact that the mean energy levels of the molecules and the suspended bodies must balance, that the latter will move around at a speed of about four millimeters per second. Unfortunately, this is still about 10,000 times faster than what was being observed.

Here Smoluchowski comes to the crux of the matter. The impacts of the molecules induce the suspended body to describe zigzag movements, and the body may indeed move with a speed of about four millimeters per second along the zigs and the zags. However, with 10^{20} molecules hammering into the body every second, the latter constantly changes direction. Hence the individual zigs and zags may well add up to four millimeters per second, but they are so tiny that they cannot be discerned, even under a microscope. As a result of very many zigs and very many zags a barely perceptible shift does occur in the body's position. It is this displacement—about one-twentieth of a millimeter per hour—which scientists can observe. Phew!

Smoluchowski noted, as an aside, that finer measurements of the Brownian motions lead to longer observed paths. Obviously, when measurements are more precise, more zigs and more zags can be measured. Furthermore, in the same manner as the displacement of a body suspended in liquid is proportional to the square root of the elapsed time, the increase in the path's length due to finer measurements is proportional to the square root of the measurements' accuracy. For example, if the position of the suspended bodies is observed ten times per second, Smoluchowski wrote, the length of the displacements will be $\sqrt{10}$ times longer than if observations take place only once every second. With this, Smoluchowski anticipated the notion of fractals, which were "officially" discovered sixty years later by the French mathematician Benoît Mandelbrot. Mandelbrot noted, for example, that on geographical maps the coast of England seems to become longer the finer the scale; small inlets, coves, and bays only become visible when the scale is increased. The coast of England and the zigzag movements of Brownian motion are fractals.[8]

To summarize, the observed jittery motions are the displacements due to the unobservable zigzag movements, which are the result of a huge number of impacts. The numerical computations that follow from the theoretical considerations were reasonable ballpark figures. Thus both Smoluchowski and Einstein had given convincing indications of the existence of molecules. This is why the notable German physicist Arnold Sommerfeld said of Smoluchowski that "his name will, forever, be associated with the first flowering of atomic theory." At the centenary of Smoluchowski's article on Brownian motion, a conference paper in his honor stated that "his name appears together with the names of Maxwell, Boltzmann and Einstein in the history of creative applications of the theory of probability to the description of physical phenomena."

There was one more contemporary theoretical explanation of Brownian motion. It was made in 1908 by the French physicist Paul Langevin. Somewhat simpler and more direct than Einstein's and Smoluchowski's, it is the one that is usually presented in today's textbooks. By the way, Langevin has another claim to fame apart from being one of the early physicists to work on Brownian motion, relativity theory, and the piezoelectric effect.[9] He was the sometime lover of the most famous woman scientist of all time, Marie Curie. In the interest of science some of the pertinent details of this affair must be revealed.

When Madame Curie's husband, colleague, and cowinner of the Nobel Prize, Pierre Curie, died as the result of a traffic accident in 1906, Marie Curie was only thirty-eight years old. The widow was devastated but eventually fell in love with a brilliant former pupil of her husband's. Paul Langevin was unhappily married; his wife reproached him for devoting his life to science instead of getting a regular job in industry that would allow him to keep her and the four children in more comfortable circumstances. Not surprisingly, he was drawn to Madame Curie. Unfortunately, Madame Langevin intercepted letters between the lovers and presented them in divorce court.[10] Many of Curie's colleagues and much of the public turned against her, accusing her of being a home wrecker. Rumors of her being Jewish, which had first appeared when she was the candidate for a seat in the Académie des Sciences—she had lost out in a scandalous vote to a Catholic rival—resurfaced and were bandied about by right-wing elements. Protests in front of her home and lab followed, so that she and her daughters had to take refuge at a friend's house. A pistol duel between Langevin and a journalist fortunately ended without shots being fired, and another one, this one with swords, between two writers ended when one of the duelists was wounded in the arm and wrist. When the Nobel Committee informed Madame Curie that she would again be the recipient of a Nobel Prize, this time in chemistry, committee members tried to dissuade her from attending the awards ceremony. Curie defied the insolent request, pointing out that "there is no connection between my scientific work and the facts of private life." An interesting postscript to the story is that Madame Curie's granddaughter and Monsieur Langevin's grandson, both nuclear physicists at the French Institut de Physique Nucléaire, are married to each other.

In the lead-up to World War II, Langevin became a vocal antifascist and peace activist. He had toyed with Marxism ever since his student days, and finally joined the French Communist party. After the Nazis invaded France in 1940, Langevin was imprisoned for a short time by the Vichy government. He escaped to Switzerland and returned to France after the war. He died in 1946.

Back now to Brownian motion. While Einstein had looked at a collection of bodies immersed in fluid, Langevin considered what happens to the individual body. Bodies in fluid are subject to two forces—the molecular kicks that cause the Brownian motion and a force that resists movement due to viscous friction in the liquid. But the viscous force that a Brownian particle experiences from the fluid is only an average. Langevin also took into account the fluctuations about this mean value which come about from the random bombardment that the suspended bodies incur. He thus added an additional random force to the equation of motion, solved it, and received the same relation as Einstein for the displacement of a Brownian particle: the average displacement—defined as the square root of the mean of the squared displacements—varies with the square root of time.[11]

In spite of compelling arguments made by Einstein, Smoluchowski, and Langevin, the evidence for the molecular structure of matter was still circumstantial. Formally it was no more than a hypothesis in need of a proof. Only experimental verification of their theoretical predictions (e.g., that the displacement of bodies does indeed vary with the square root of time) would put the doubters to rest. It would fall upon a friend of Langevin's, Jean Perrin, to provide the missing link.

But before Perrin started his work, someone else was already at it. Theodor Svedberg was a twenty-two-year-old student of chemistry in Sweden when he started his career as an experimentalist. "The," as he was known among his friends—to the general confusion of readers who stumble at mentions of The Svedberg—was born in 1884. His father, a civil engineer, managed ironworks in Sweden and Norway. He loved the outdoors and often took his only son on hikes in the countryside, nurturing in him an appreciation for nature, especially botany. His high school teachers noticed the boy's talents and let The use the physics and chemistry labs after school hours.[12] These extracurricular experiments made him decide to study chemistry in spite of an intense interest in botany. He was convinced that chemistry contained the answer to many unsolved questions in biology. The entered the University of Uppsala at age nineteen and then hit the fast track. His professors gave him credit for the experiments he had done during high school, and the Svedberg finished his undergraduate studies in twenty months. Armed with a Fil. kand. degree (B.A.), he felt ready to conduct real experiments.

His academic career got off to an ignominious start, however. Maybe it was a feeling of internal kinship with botanist Robert Brown that induced The Svedberg to start his work with experiments on Brownian motion. In 1906, a year before he was awarded his doctorate, he used an ultramicroscope to peer at little bodies suspended in liquid. The instrument had been invented only a few years earlier by Richard Zsigmondy, an Austrian chemistry professor, and Henry Siedentopf, an optician with Carl Zeiss, the optical instruments manufacturer. By scattering light, it made particles visible that were too small to be perceived by the ordinary microscope. Zsigmondy had already witnessed Brownian motion and described it as "a swarm of dancing gnats in a sunbeam . . . They hop, dance, jump, dash together and fly away from each other, so that it is difficult in the whirl to get one's bearings." Zsigmondy believed that he could distinguish a translatory movement and, superimposed on it, a fast oscillatory movement. Strongly influenced by Zsigmondy's findings, Svedberg believed that he too observed the oscillations. Furthermore, he thought that he had managed to measure the bodies' velocity. The young man submitted his findings to the *Zeitschrift für Elektrochemie*, where they were duly published.

He was wrong on both counts. Measuring velocities was a hopeless undertaking, as Smoluchowski pointed out at about the same time.[13] And oscillations . . . well, they were nowhere to be seen. Actually, Svedberg's experiment was quite clever. He let the fluid containing the suspended particles flow with a known velocity in one direction under the microscope. This would pull the oscillations—if there were any—apart, and the experimenter should observe a sinusoidal movement. Svedberg convinced himself that he did see undulating motions in the fluid and even estimated their amplitude, frequency, and velocity.

Unfortunately, he was led astray by his wish to observe what he expected to observe. Svedberg had made all estimations by visual inspection, which is a dangerous form of measurement, especially when the experimenter has a preconceived opinion about the expected results. Seventy years later, an American chemistry professor reviewed the historical facts and commented that Svedberg had made a number of errors, employed curious reasoning, piled confusion upon confusion, and performed meaningless mathematical manipulations. "It would be best to dismiss [the paper] as an immature and fanciful effort of an imaginative and resourceful young student," he concluded.[14]

But Svedberg noticed none of that. Proudly he sent the paper to Einstein, thinking that he would be happy to see his theoretical predictions proved experimentally. Einstein was underwhelmed. In fact, he immediately sent a note to the *Zeitschrift für Elektrochemie* pointing out some of the errors in Svedberg's article. In order not to hurt the young student's feelings, whose obvious enthusiasm for research he did not want to stifle, he couched his criticism in gentle words. "I would like to point out some properties of [Brownian] motion indicated by the molecular theory of heat. I hope I will be able by the following to help physicists who handle the subject experimentally, interpret their observations."

In spite of the careful language, the bottom line was inescapable. Since the individual legs of the particles' paths, which themselves make up the zigs and the zags of Brownian motion, are so minute, it is impossible to measure the velocity of the suspended bodies. Svedberg obviously misinterpreted what he believed he saw. "The mistakes in Svedberg's methods of observation and also in the theoretical methods of treatment had become clear to me at once," Einstein wrote in a letter to Jean Perrin. "The velocity [that Svedberg purported to measure] corresponds to no objective property of the motion under investigation." In other words, Svedberg measured some virtual quantity that did not mean anything. Thus it was impossible to verify Einstein's alleged prediction that the root mean square velocity of the suspended particles is proportional to the square root of time. (What is observable, and what Einstein actually predicted, is that the root mean square displacement is proportional to the square root of time.)

Unlike Einstein, Perrin felt no compunction to mince words. "Questionable methods," "vague estimates," "ill-defined magnitudes" were just some of the epithets he used. In particular, he charged, when Svedberg claimed that he had observed oscillations he was "evidently victim of an illusion." Svedberg persisted with the assertion that he had observed oscillations. If one observes a particle in a still fluid, he wrote, "one gets the general impression of a hither and thither movement." Well, as justifications go, this was a rather weak one.

Svedberg never admitted to the shortcomings of his first experiment conducted in the bloom of his youth. Years later, when he was a famous scientist, he still insisted on the correctness of his observations and interpretations and presented illustrations that purportedly show the oscillations. The pictures are actually quite embarrassing, because even a layman can ascertain that oscillations are nowhere to be seen. Svedberg sometimes also claimed that he had anticipated Einstein's findings because he wrote his first paper before being aware of the original article of 1905.[15]

After the experiments on Brownian motion, The Svedberg went on to do really path-breaking work. For example, he invented the so-called ultracentrifuge, a machine that is capable of rotating 120,000 times per minute, thus producing a force equal to half a million times the earth's gravity. With the help of this new device, substances of greater and lesser density in a suspension, or molecules in a solution, could be separated. It enabled scientists to discern the size and shape of certain particles. It also allowed biologists, biochemists, physicians, and other life scientists to inspect viruses and examine how they attack cells. Today Svedberg is rightly considered the founder of molecular biology.

Svedberg remained at the University of Uppsala all his life, except for an eight-month sabbatical at the University of Wisconsin in 1923. Married four times, he had a dozen children, six boys and six girls. He was honored for his work with honorary doctorates from the Universities of Oxford, Paris, Delaware, Groningen, Uppsala, Wisconsin, and Harvard. And in 1926, he silenced those who still doubted his achievements by winning the Nobel Prize. Somewhat to his embarrassment, his research on Brownian motion figured prominently in the laudation. His selection for this most prestigious of all scientific awards was preceded by some dramatic events, as described below.

Let us return to Paris. Jean Perrin and Paul Langevin had been close friends since their student days, when they not only studied physics together at the École Normale Supérieure but also discussed socialism and Marxist philosophy. They were part of an intimate group of Paris Left Bank intellectuals who believed that revolutionary Marxist ideas complemented the revolution in physics that was taking place at the turn of the century. They even bought shares in the Societé Nouvelle de Librairie et d'Edition, a company that edited and distributed socialist literature. Apart from

Langevin, some of Perrin's closest friends were Marie and Pierre Curie and their daughter and son-in-law Irène and Frédéric Joliot-Curie (joint Nobel Prizes in chemistry, 1935). The Curie and Perrin families lived in neighboring buildings on Boulevard Kellermann in Paris's thirteenth arrondissment and often met for Sunday dinners, regularly joined by other physicists, chemists, and scholars from different fields. For example, the mathematician Émile Borel and his wife, the writer Camille Marbo—daughter of the noted mathematician Paul Appell, member of Louis Bachelier's thesis committee—were frequent guests. (Camille and Émile were the ones who would shelter Marie Curie and her daughters in their home during the Langevin affair.).

Perrin was born in the French town of Lille in 1870. When Jean was still a baby, his father, an infantry captain in the French army, was fatally wounded in the Franco-Prussian War. Eventually the talented boy was sent to high school in Paris. After graduating, he received intensive mathematics instruction for two years, in preparation for the competition to enter the École Normale Supérieure, the indispensable springboard for academic careers in France. Of course he passed the exam and in 1891 entered Normale Sup. There he not only studied mathematics and physics but also fell under the spell of Lucien Herr, Normale Sup's legendary librarian, one of the first Frenchmen to study the writings of Karl Marx in depth. As the Dreyfus affair rocked France, Herr made sure that Normale Sup was a stronghold for Dreyfusards, the supporters of the wrongly accused army captain.

After graduating, Perrin became an assistant at the school and began work on his Ph.D. But even before his thesis was finished he made waves. In 1895 he twenty-five-year-old doctoral student showed that cathode rays are not light-like vibrations of an imagined ether, as many believed, but are actually negatively charged particles that we today know to be electrons. In 1910 Perrin became professor of physical chemistry at the Sorbonne, where he spent most of his career. When World War I broke out, Perrin, like many left-wing intellectuals, crossed over seamlessly from the antimilitarism of the post-Dreyfus period to the patriotism that was prevalent in 1914. Swept up by nationalism, most French leftists joined the effort to defend their fatherland against the German Empire. Marie Curie and her teenage daughter Irène, for example, traveled with X-ray equipment to the front to assist the wounded. Perrin served as an officer in the French engineering corps, working with Langevin and others on ballistics and on the detection of airplanes and submarines by sound. The war was devastating for French intellectuals. Of the 161 students who entered Normale Sup between 1911 and 1913, nearly 90 percent were killed or wounded.

In 1936 Perrin entered politics. He participated in Leon Blum's Popular Front cabinet, a coalition of left and center parties, becoming undersecretary of state for scientific research. "He appeared naive and distracted, almost in the clouds," Jean Zay, the minister of education, remarked, "but in reality he was always attentive, precise, concentrated, even crafty if it was necessary." As undersecretary he founded the Centre National de la Recherche Scientifique (CNRS), the organization that is also today the center of French science, and—ever mindful of the proletariat—the Palais de la Découverte, a science museum first provisionally constructed for the Paris Exposition of 1937 and still one of the most popular museums of Paris.

When the Nazis invaded France in 1940, Perrin wanted to join the resistance, but friends thought he would do more good abroad as a spokesman for the cause of a free France. They persuaded him to join his son Francis who was teaching at Columbia University. Perrin followed their advice and spent his time in New York writing articles and attending meetings and conferences. He helped found the École Libre des Hautes Études, a sort of university-in-exile for French expatriate intellectuals. Most of the time he was in poor health and he died in 1942, while his beloved France was still under occupation. After the liberation, his remains were repatriated and interred in the Panthéon in Paris.

Throughout his career, Perrin was showered with honors. He was a member of the Royal Society of London and of the Academies of Sciences of Belgium, Sweden, Turin, Prague, Rumania, and China. In 1923 he was elected to the French Academy of Sciences. He held honorary doctorates of the Universities of Brussels, Liege, Ghent, Calcutta, New York, Princeton,

Manchester, and Oxford. He became a Commander of the Legion of Honor in 1926, and was also made Commander of the British Empire and of the Belgian Order of Leopold. And in 1926, when Svedberg received the Nobel Prize in chemistry, Perrin received the same honor in physics "for his work on the discontinuous structure of matter." By "discontinuous structure" the Nobel Prize Committee meant that matter is made up of atoms and molecules, with voids in between.

In his Nobel speech Perrin explained. "A fluid such as air or water seems to us at first glance to be perfectly homogeneous and continuous.... However, this can be taken for granted only up to the degree of subtlety reached by the resolving power of our senses.... in order really to establish the Atomic Theory, it was necessary to obtain the weights and dimensions of the atoms." Hence, if the size of these until-then only imagined building blocks could be established, doubters of the atomic theory would be hard-pressed to argue that they did not exist.

Even though Perrin could not yet determine the atoms' size, he did something equally cool. He determined Avogadro's number, the number of atoms in 12 grams of carbon-12, or in the equivalent mass of some other chemical element. His numerous experiments yielded an Avogadro number of between 6.0×10^{23} and 6.8×10^{23}. (Recall that Einstein had estimated it to be 4×10^{23}, which is very close in terms of order of magnitude.) With the number of atoms that are present in a mass of material determined, their actual existence was a foregone conclusion. Perrin summarized that "the objective reality of molecules and atoms which was doubted twenty years ago, can today be accepted as a principle, the consequences of which can always be proved."

Like Einstein, Perrin also harbored a hope. "It would still be a great step forward in our knowledge of matter ... if we could perceive directly these molecules, the existence of which has been demonstrated." That would still take a while, however. It was only in the 1950s that the German physicist Ernst Müller invented the field ion microscope, which had a magnification of about 10 million times. It provided the first clear view of crystals on an atomic scale, showing the individual atoms and their surface arrangement. But already in the 1910s, few people still doubted the reality of atoms. The truth of Einstein's, Smoluchowski's, and Langevin's theories had been established.

Perrin's experiments were a model of precision and patience. Since the rate of diffusion depends on the size of the particles, his first task was to isolate particles that had the same size, not a simple undertaking with particles that measure one-thousandth of a millimeter across. Using gamboge, the resin of a tree that Rembrandt used as a yellowish pigment for his oil paintings, Perrin and his students were able to isolate similar-size particles by ingenious use of multiple centrifugations. It was painstaking work. Starting with 1.2 kilogram they managed to separate a few tenths of a gram of suitable material. And after these arduous preparations, the work had only just begun. Extensive series of experiments involving the counting of many thousands of particles were necessary to enable good statistical estimates.

Perrin reported on his first experiments to the Académie des Sciences on May 11, 1908. It was exactly one week before the polymath Victor Henri—who at different times was active in physics, chemistry, biology, and diplomacy (he was the French ambassador to Russia during World War I)—reported on his cinematographic studies of Brownian motion before the very same assembly. His photographic evidence, shot at intervals of one-twentieth of a second, confirmed qualitatively, but contradicted quantitatively, the predictions of Einstein, Langevin, and Smoluchowski.[16] The reason for the numerical discrepancies is unknown, but later physicists assumed that it may have been the strong illumination that the then new technique of photography required, which may have heated the fluid more than expected.[17]

Perrin's mode of attack was to let the gamboge particles diffuse under gravity. (Particles heavier than the liquid in which they are suspended perform the jig also on their way down.) Like Svedberg, he first tried to measure the speed of the suspended particles, but soon realized that that was the wrong approach. "Previously we were obliged to determine the 'mean velocity of agitation' by following as closely as possible the path of a grain," he wrote. "Values so obtained were always a few microns per second for grains of the order of a micron. But such evaluations of the activity are absolutely wrong." Perrin had come to understand that speeds seem to get higher as the measurements get finer because—recall Mandelbrot's fractals—the convoluted microscopic paths seem to become longer with greater magnification. So he decided to concentrate on observing not how fast the particles move, but how they are distributed after a while.

At the outset of the actual experiment, he let the suspension settle for a few hours, supposing that gravity, acting downward, and osmotic pressure, acting upward, would eventually balance each other. He was right; in due course gamboge particles arrived at a steady state. Perrin and his students then counted the particles meticulously at various heights (varying the observed height level by moving the microscope's objective lens up or down). They did this not once or twice or thrice . . . no, they repeated the procedure several thousand times (!) and only then computed the averages. In that manner, Perrin observed that the gamboge particles were distributed in the fluid according to the exponential distribution. This is what also happens with particles in the atmosphere, except that what takes place in the sky over a height difference of 6,000 meters occurred in Perrin's suspension in one-tenth of a millimeter. By measuring displacements, rather than zigzag distances traveled, the experiments yielded, albeit vertically, what Einstein, Smoluchowski, and Langevin had predicted: the square root of time law.

In 1909 Perrin published a 110-page report of his results in the *Annales de Chimie et de Physique*. Entitled "Mouvement Brownien et la réalité moleculaire" (Brownian motion and the molecular reality), it offered observations on Brownian motion as irrefutable proof of the existence of molecules. Perrin constantly refined his techniques and repeated his experiments with different materials suspended in different fluids with differing viscosities, under various conditions. He was the first to truly confirm Einstein and Smoluchowski's theoretical predictions, thus raising them from a hypothesis to a description of matter. In the end, even arch-skeptics of the molecular hypothesis were convinced. Einstein himself was full of praise. "I would have thought it impossible to investigate Brownian motion so precisely," he wrote in a letter to Perrin. "It is a stroke of luck for the subject that you took it up." In 1913 Perrin gave an account of the search for the true composition of matter in *Les Atomes,* a book for the general public that became a best-seller in its time and is still in print nearly a century later. As a tragic footnote, many of Perrin's students and coworkers lost their lives on the battlefields of World War I.

As noted above, the Swedish Academy of Sciences awarded Jean Perrin the Nobel Prize in physics in 1926. The Svedberg received the chemistry prize in the same year "for his work on disperse systems." And since the chemistry prize for the previous year was awarded to Richard Zsigmondy "for his demonstration of the heterogeneous nature of colloid solutions" and the physics prize to James Franck and Gustav Hertz "for their discovery of the laws governing the impact of an electron upon an atom," there were four Nobel Prizes in two years that dealt with the molecular structure of matter. Thus it was finally accepted by everyone, scientists and laypeople alike, that matter is made of molecules and atoms.

The selection of the winners by the Royal Academy of Sciences had not been quite straightforward. For reasons that are no longer clear, both the 1925 and the 1926 prizes were announced in November 1926. They were awarded jointly a month later, on December 10, the anniversary of Alfred Nobel's birthday. Nowadays, the Nobel committee keeps details about the internal deliberations secret. But in a series of unpublished autobiographical notes, Svedberg allowed a glimpse into the lowdown of the selection process. As a member of the Royal Academy of Sciences, he was privy to the deliberations of this august body. [18]

In the early years of the Nobel Prize, the awards did not always recognize brilliant achievements as we would like to believe, and the deliberations of the committees were not as arm's length as we hope they are today. Rather, they reflected "changing priorities, arrogance, racism, hostility, sexism, inconsistencies, politics, ambitions, open and hidden agendas, biases, rivalries, vanities, pettiness, prejudices, and narrow personal, scientific and cultural self-interests of committee members . . . On several occasions the committee denied recognition of true brilliance while rewarding mediocrity."[19]

Usually the winners were selected in November. Committees put forward the names of the chosen candidates in physics and in chemistry, and in a plenary meeting the full academy had to decide on the winner. The committee's choice was just a recommendation; the academy was not bound to accept, or even consider, the committee's preferred candidate. Traditionally, the physics prize was decided first and the choice of Franck and Hertz for the 1925 prize was accepted without great opposition. But the committee's choice for the 1926 prize was another matter. The academy was dissatisfied with the committee's suggestion (it is not known who its intended prize winner was) and, in direct opposition to the recommendation, the nod went to Jean Perrin, who had been suggested by the world-famous metallurgist Carl Benedicks of the University of Stockholm.

When the time came to choose the recipients of the chemistry prizes, things disintegrated. Due to his work on the ultracentrifuge, Svedberg had a solid reputation, enjoyed much support among his peers, and was certainly a contender for the prize. But in order to be considered, he would have to excuse himself from the chemistry committee. Since he considered his chances rather slim, he declined a nomination and remained on the committee. After deliberations, the committee decided that Richard Zsigmondy should be put forward for the 1925 prize, and the committee's recommendation was accepted by the academy. So far so good.

Now it was time to decide on the 1926 prize. Support for Svedberg had grown among the academy members while the 1925 prize was being discussed. Again it was Carl Benedicks, together with another colleague, who led the fight for their preferred candidate. What a shame to let this chance pass him by, Svedberg must have thought to himself, and to the dismay of the committee chairman, who had already set his eye on a different candidate, he decided to leave the meeting so that he could be considered for the prize.

Olof Hammarsten, professor of medical and physiological chemistry at the University of Uppsala, was chairman of the chemistry committee. At eighty-five, he was the leading biochemist in Sweden. Carrying great weight among his colleagues, he was used to getting his way at the university as well as in the Nobel committee. And he was upset with this obstreperous committee member. "But you already declared in the committee that you do not wish to be considered," he boomed. Svedberg was not intimidated. "In order to be able to take part in the work of the committee, I asked that it should not take my candidacy into consideration," he retorted, "but if the academy now wishes to do so, I have no reason to declare that I cannot come under consideration." And with that he walked out.

While pacing the platform of the Central Station, waiting for his train home, acquaintances approached him with the good news. After he left, the academy had indeed voted to award him the 1926 Nobel Prize for chemistry. "I was overwhelmed, I had not thought at all that it would go this way," he confided in his notes, sounding a wee bit insincere, what with his having provoked a clash with the committee chairman over his candidacy. Feelings of discomfort started to arise. Svedberg felt deep down that his life's work did not merit such an honor. Maybe his newest endeavor, his work on proteins in the ultracentrifuge, would eventually warrant a Nobel Prize, but this work had just begun. Maybe the committee had been right—he should have waited a few more years before proffering his candidacy. All night he lay in bed brooding. Finally he put himself somewhat at ease: he would devote the following ten years of his life to making himself worthy of the prize.

Not everybody understood why Svedberg had been chosen by the Nobel committee, and it was generally assumed that the academy had seen publications of his new research on proteins before they became generally known abroad, thus justifying the choice. It was then decided that even though the awards ceremony would be on December 10, Svedberg's prize lecture would be postponed to the following spring. The unofficial reason was that the Nobel committee should not feel the humiliation of having been ignored so bitterly. Even more unofficially, but probably more to the point, it was said that the five-month postponement would enable Svedberg to collect more material from his protein investigations.

To Svedberg's chagrin, when the secretary of the Royal Swedish Academy of Sciences gave the laudation speech at the awards ceremony, he dwelt at length on Svedberg's early work on Brownian motion . . . which had been known to be faulty for nearly twenty years. His language was very descriptive. "If a fly or a gnat flies against an elephant, the elephant will not noticeably alter its position, but this can occur if the fly or gnat collides only with a bee," the secretary expounded. "Should it now be true that the movement of particles suspended in a liquid . . . can be explained only as a result of the movement of molecules . . . then this provides visual evidence for the real existence of molecules and consequently also for that of atoms, evidence which is all the more remarkable as not so long ago an influential school of scientists declared these particles of matter to be unreal fictions representing an obsolete viewpoint of science." As the ceremony's main speaker, he then took great pains in emphasizing the role that the two prize winners, Perrin and Svedberg, had played in experimentally confirming Einstein and Smoluchowski's theory. Perrin must have cringed.

When the time finally came, in May 1926, for Svedberg to give his prize lecture, he all but forgot to talk about Brownian motion, mentioning it only in passing in the context of the measurement of particle size. The royal dinner was dull and rather cumbersome. "The king, Gustav V, was served first, ate little and rapidly, and those who got food last, scarcely had time to eat it," Svedberg recounted. The dinner party Carl Benedicks gave a few days later for Perrin, Zsigmondy, and Svedberg was much more relaxed. In his after-dinner speech he compared the prize winners with the three wise men. "I was Balthazar," Svedberg recalled. The Nobel Prize strengthened Svedberg's position at the University of Uppsala—a new building for physical chemistry was soon built—and the prize money meant a welcome windfall for him, his wife, his ex-wife (to be followed by two more ex-wives), and eventually twelve children.

As we will see in the following chapters, the square root law, which says that the displacement of a particle after a certain time is equal to the square root of the elapsed time, is pertinent indeed for the theory of finance.

SEVEN

Disco Dancers and Strobe Lights

S OMETIMES A PROBLEM THAT PREOCCUPIES SCIENTISTS IN
one discipline may pop up in another. Lucky latecomers may
then be able to make use of the tools that have already been developed by their colleagues. In turn, their attempts to tackle their own
problem may shed a new light on the previously studied phenomena.
Thus it was with random movement. First discovered by biologists, the
phenomenon was subsequently studied by physicists and chemists and,
in due course, by mathematicians and statisticians. Eventually, as will be
recounted in later chapters, the subject would also become important in
economics and finance. So, while theoreticians in Germany and Poland
were getting a grip on the concept of Brownian motion, and experimen-
talists in Sweden and France were busy demonstrating its molecular origin,
a scientist in England was wrestling with the same question in a different
context. On July 27, 1905, just eleven weeks after Einstein submitted his

paper to the *Annalen der Physik,* a letter from a reader was published in the science magazine *Nature* asking for help with a certain question.

The letter was written by Karl Pearson, an English polymath well versed in psychology, history, German literature, law, and mathematics. It appeared under the subheading "The Problem of the Random Walk." It read, "Can any of your readers refer me to a work wherein I should find a solution of the following problem, or failing the knowledge of any existing solution, provide me with an original one? . . . A man starts from the point O and walks z yards in a straight line, he then turns to any angle whatever and walks another x yards in a second straight line. He repeats this process n times. I require the probability that after n stretches he is at a distance between r and $r + \delta r$ from his starting point O." The question was motivated by Pearson's interest in the spread of disease. A few years earlier, Major Ronald Ross, a doctor in the Indian Medical Service, had identified the mosquito as the source of malaria. (For this achievement Ross was awarded the Nobel Prize for medicine in 1902.) Subsequently Pearson studied how these insects invade a cleared jungle region. Their buzzing about until they eventually take over the entire area was described by him as a "random walk." [1]

Pearson, who would become known as the father of modern statistics and also, sadly, as an advocate of eugenics, did not have to wait long; the answer would appear in the following week's issue, signed simply "Rayleigh, Terling Place." *Nature* readers knew who was behind the terse signature; the respondent was none other than John William Strutt, third Baron Rayleigh, who had received the Nobel Prize in physics just a year earlier. Rayleigh's answer referred Pearson to work he had done on vibrating strings more than a quarter century earlier. The results had been published in *Philosophical Magazine* in 1880.

Among the questions Rayleigh investigated was, "What we are to expect from the composition of a large number (n) of equal vibrations of amplitude unity, of the same period, and of phases accidentally determined?" What amplitude does a string have if vibrations are superimposed? He first pointed out that two vibrations with the same period and amplitude, but with opposite phases (i.e., offset by one-half of the period) cancel each other. When one vibration points upward, the other points downward, and the string does not vibrate at all. It is the drunken sailor all over again: a step forward and a step backward annul each other, and the drunkard ends up at the lamppost as if he had never left it.[2]

With this groundwork out of the way, Rayleigh went on to analyze what happens when more than two, in fact when many vibrations are superimposed and their phases are randomly distributed either up or down. How does the amplitude of the vibration change? He found that when the number of vibrations superimposed on each other increases toward infinity, the probability of the amplitude reaching a certain level corresponds to the normal distribution. Lo and behold, this was exactly the same thing that would be derived by Einstein and Smoluchowski twenty-five years later. They would show that when molecules hit the particle from random directions, the probability of a certain displacement of the particle would be governed by the normal distribution. Furthermore, Rayleigh showed that the amplitude of the superimposed vibrations varies, on average, with the square root of the number of vibrations. This was also the same result that Einstein and Smoluchowski would find. In a totally different context, Rayleigh's computations had preceded theirs. Apparently there is nothing new under the sun; drunken sailors, vibrations, and particles all behave in the same manner. Their lurching, amplitude, and traveled distances increase with the square root of steps, vibrations, and time, respectively.

Sensing that some people might get the wrong idea, Rayleigh was quick to add a caveat: "The reader must be on his guard here against a fallacy which has misled some eminent authors. We have not proved that when n [the number of vibrations] is large, there is a tendency for a single combination [of vibrations] to give an intensity equal to n, but the quite different proposition that in a large number of trials, in each of which the phases are distributed at random, the mean intensity will tend more and more to the value n." ("Intensity" in the physics of sound is the square of the vibration's amplitude, so when Rayleigh says that the intensity is equal to n, he means the same as "the amplitude grows with the square root of n.")

To readers educated in mathematical analysis but unfamiliar with statistical methods, there was indeed the danger of error. Mathematicians were used to a function tending toward a certain value, as its variable grows to infinity. But in statistics, on any specific trial or experiment, the function need not actually hit that value. Only the mean of many trials would tend toward that specific value. For example, when a series of twenty coins are tossed, on average ten of them will come up heads. But this is true only on average; it may not actually happen on any specific throw. In fact, a throw of twenty coins may result in twelve heads, another in four heads, and so on. When the twenty coins are thrown a large number of times, however, the average number of heads will become ever closer to ten. Similarly, none of the drunken sailors may actually move a distance that corresponds to the square root of his steps. But if we have many drunken sailors, they will, on average, move by that much.

Another week later, Pearson thanked readers who had answered his letter and apologized to Rayleigh for not having read his treatise before. "I ought to have known it but my reading of late years has drifted in other channels." He justifies himself by remarking that "one does not expect to find the first stage in a biometric problem provided in a memoir on sound" . . . and not in *Philosophical Magazine*, one might add. The sequence of events illustrates that researchers never know what to expect from their solutions in mathematics. What is true of vibrating strings may be true of particles suspended in a fluid, drunks tottering around a lamppost, and even financial instruments traded on the stock market, as we will see in the coming chapters. Applications may appear in unexpected contexts.

To end the exchange of letters, Pearson announces the bottom line. "The lesson of Lord Rayleigh's solution," he concludes, "is that in open country the most probable place of finding a drunken man who is at all capable of keeping on his feet is somewhere near his starting point." Back to the lamppost it is.

Pearson was right to stress "in open country," even though he did not know it at the time. Many years later, it was proved that random walks in one or two dimensions (i.e., along a line or in a plane) always return to the origin as the number of steps grows to infinity. Amazingly, this is not true in three dimensions. The probability of returning to the origin from wanderings in space is only about 34 percent. This prompted the Japanese mathematician Shizuo Kakutani to proclaim, "A drunk man will find his way home, but a drunk bird may get lost forever." And George Polya, a Hungarian mathematician whose book *How to Solve It* sold more than a million copies over the years, used to say that "in a plane, all ways lead to Rome." More on that later.

Did I say that there is nothing new under the sun? I did, and the more I dig, the more I find how true the saying is. In the same year that Lord Rayleigh's essay on vibrations was published in *Philosophical Magazine*, a Danish astronomer with a wide range of interests published an original but forgotten article in a journal of the Royal Danish Academy of Arts and Sciences.

Thorvald Nicolai Thiele, who contributed important work to astronomy, mathematics, statistics, and actuarial science during his lifetime, was born on Christmas Eve 1838 into a prominent Danish family. His father was the private librarian of King Christian VIII, and Thiele was raised in a stimulating environment. He studied astronomy at the University of Copenhagen, obtaining his master's degree at age twenty-two, and his doctorate six years later. He then taught at his alma mater, eventually becoming the director of the university's astronomical observatory. Apart from his teaching duties, he was a founder and the chief actuary of the first private Danish insurance company, Hafnia, for thirty-eight years and a founding member of the Danish Mathematical Society, the Danish Actuarial Society, and the Danish Chess Club. He is nowadays remembered for deriving an important differential equation that is still used by actuaries. He was also actively interested in social policy, arguing, for example, that old-age pensions should not be considered a kind of insurance, since—in contrast to disability—it is a fate that eventually befalls all people if they live long enough. Therefore pensions should be considered as social contracts between generations, with the state as mediator.

Thiele's eyesight was severely impaired. Consequently he could not conduct astronomical observations, and so he directed his interests at the observatory toward computational work. The contribution that is of greatest interest to us is a paper with the title, translated into English, "On the Application of the Method of Least Squares to Those Cases in Which a Combination of Certain Types of Inhomogeneous Random Sources of Errors Gives a 'Systematic Character' to the Errors." In this paper, published in 1880, Thiele investigated the behavior of time series—how observed variables like the movement of stars, the temperature outside, or stock prices vary over time.

In 1794 the prince of mathematics, the German Carl Friedrich Gauss, first thought of the "method of least squares" as a tool to deal with measurement errors. A bothersome fact of life for scientists was that repeated measurements of the same attribute usually give differing results. The problem was, and is, unavoidable, since measurements are always tainted by errors, whether attributable to noise in the data, inaccurate instruments, or human mistakes. To get valid estimates of the true value, the errors need to be eliminated or at least minimized. This can be achieved for a simple observation, like the weight of an object, by taking the average of multiple measurements. But if the attribute to be measured varies over time, simple averages will not do and a different method needs to be found. Let's say that observations are plotted on a piece of paper, with time as the x-axis and the measured attribute (e.g., a country's gross national product) on the y-axis. The aim usually is to draw a straight line through the data points. It expresses how the variable changes with time (e.g., how GNP grows). But how do we find the line that best fits the data? Gauss determined that a well-fitting line can be computed by minimizing the sum of the squared differences between the observed values and the values on the line. (By squaring the differences, positive and negative errors are both accounted for, instead of compensating each other.) Gauss's method of least squares is often called "linear regression" and is a standard tool in all statistical computer packages.

Thiele utilized the method of least squares to predict the true values of a process when observations are fraught with errors due to, say, the finite resolution of a telescope or a microscope. His aim was to give a model that describes the observation errors present in a sequence of measurements.

The process Thiele considered was the Brownian motion of a particle, even if he did not describe it in these terms: random steps forward and backward that, on average, balance out, and displacements that are proportional to the square root of the number of steps. Onto this process random measurement errors are superimposed. Sometimes the measurements are too large, sometimes they are too small, but on average the errors are zero. Thiele's aim was to estimate the particle's true position at various time intervals.

Starting with initial estimates of certain characteristics of the Brownian process and of the estimation errors,[3] Thiele proposed an iterative procedure to determine the true values of the particle's positions that are hidden under the noise of the measurement errors. On each iteration, the estimates became better, gradually approaching the true, if unknown, values. The procedure, precursor of the so-called Kalman filter that would be reinvented eighty years later, was implementable with tables and simple machines that add and multiply. Ease of computation was, after all, an important consideration at a time when modern-day computers were not available.

Thiele's work on the method of least squares sparkled with ingenuity. He must have been aware of that since he also published a French version of his paper. Unfortunately, its importance was not generally recognized at the time. "It is a brilliant tour de force, but so far ahead of its time that few could appreciate the results," a statistician would write a hundred years later.[4] Actually, when Pearson sent his letter to *Nature,* he should have, or at least could have, known of Thiele's work, since both men were interested in statistics. Furthermore, at least one of Thiele's students, a woman named Kirstine Smith, went on to study under Pearson. She certainly could have served as a conduit of ideas from one teacher to the other, but apparently Pearson did not bother to inquire. Thus Thiele's work went unnoted.

Within a quarter of a century Rayleigh, Thiele, Pearson, Bachelier, Einstein, Smoluchowski, and Langevin had described the Brownian process, but none of their work was rigorous by today's standards. Apart from giving a mathematical description of the jittery movements, none of them defined what a Brownian process actually was in a strict mathematical sense. The level of rigor to which scientists at the turn of the century adhered was typical for the field of physics, but to later mathematicians their treatment of the Brownian process was just a lot of hand waving. It was left to MIT professor Norbert Wiener to fill the gap. In his honor, the mathematical model of the physical Brownian motion is often called the Wiener process.

Wiener was an outstanding mathematician whose name, but not his work, is widely overlooked by the general public. He was a child prodigy, graduating from high school at eleven and from college at fourteen, and obtaining his doctorate at eighteen. Sadly, Wiener came by his brilliance the hard way. Leo Wiener, the intelligent but imposing father, had taken over the boy's education. The renowned, self-educated Polish immigrant taught Slavic languages and literature at Harvard but was also conversant in philology as well as most subjects in higher learning, including mathematics. His manner of instructing his son consisted of a rigorous regimen of home schooling, with humiliation at the slightest error. Claiming that the key to his son's success was not the son's talent but his own unique manner of educating the supposedly good-for-nothing boy, the father's favorite educational technique was to insult him as an ignorant blockhead. This extraordinary method did produce the desired results but at a heavy price: throughout his life, Norbert Wiener suffered from depression. Eventually Wiener seems to have made his peace with his father, calling him "my closest mentor and dearest antagonist" in the preface to one of his books.

After he got his bachelor's degree in mathematics from Tufts College, the fourteen-year-old boy entered Harvard to begin graduate studies in zoology. But the subject did not satisfy him and he changed to philosophy. His doctorate in mathematical logic was followed by visits to Cambridge, England, where he studied with Bertrand Russell and G. H. Hardy, and to the University of Göttingen in Germany, the high temple of mathematics at the time, where he met David Hilbert and Edmund Landau. Back in the United States, he taught philosophy at Harvard but could not get a permanent position because of a quota that restricted the number of Jewish faculty members.[5]

As a child Wiener had not even known that he was Jewish. His parents had kept his ethnic background a secret and his mother was not above making the occasional anti-Semitic remark. All the greater was the boy's shock when he found out about his ancestry at age fifteen. Now he also had to deal with an identity crisis. A few years later, he was not allowed to marry his first love, an astrophysicist, but had his wife chosen for him by his parents. Margaret Engelmann, a gentile girl of German descent, was one of his father's students. The parents thought her well-suited to take over their duties of caring for the awkward young man who lacked any social skills. Unfortunately, the woman developed an extreme form of jealousy that distanced Wiener from his colleagues. Apart from furthering her husband's career by sheltering him from the insidious aspects of everyday life, her intrigues and her meddling in the working relationships with his collaborators meant no end of trouble for him. She may even be the reason why Norbert Wiener never became a household name. To top things off, she became a fervent admirer of Adolf Hitler's, keeping a copy of *Mein Kampf* on her night table.

Wiener's work spanned pure mathematics, applied mathematics, electronic engineering, computer science, communications theory, and other areas. He created the field of "cybernetics" (a term he himself coined), which is a statistical approach to the theory of communication. Motivated by government work during World War II on antiaircraft guns, Wiener became interested in automation and control theory, a branch of engineering that deals with the behavior of machines, robots, vehicles, weapons, and other dynamical systems. The objective was to make antiaircraft fire more effective. The operator of an antiaircraft gun would track an enemy plane for about ten seconds—by sight in World War I, by radar in World War II—predict its location another ten seconds hence, aim his gun, and fire toward that point. Key to a successful kill was the operator's stable hand, steadied by his own body's internal feedback mechanism—his ability to avoid jitter and continually adjust his aim for minute errors. Feedback, the all-important tool in the gun operator's bag of tricks, would become one of the hallmarks of cybernetics, the science of how to control and regulate machines and organisms. Wiener's task was to automate the tracking, aiming, and firing routine. Basically, the information the operator receives is a stream of data from which he must filter away the noise so that the signal, in this case the aircraft's flight path, can be identified. It is the same problem that communication engineers confront when trying to make messages intelligible against background noise.

Random noise and the probabilities associated with it brought to Wiener's mind early work he had done on Brownian motion. When he started working on the antiaircraft tracking and aiming system, Wiener had been familiar with Brownian motion for nearly thirty years. When Wiener had visited Cambridge University in 1914, Bertrand Russell commended Albert Einstein's paper to the twenty-year-old postdoctoral student. As a mathematics instructor at MIT, Wiener started dealing with the theoretical aspects of Brownian motion. He was the first to put the mathematical treatment of Brownian motion on a rigorous footing. Instead of simply describing the movement as jerky or jittery, he identified four clear-cut properties that a dynamic process must possess in order to qualify as Brownian motion. First, at time zero, the process starts at some point zero, but that is only a convention, since—to belabor the drunken sailor paradigm—the lamppost where the drunken sailor starts his walk could be anywhere. Second, the steps must be independent of each other. This means that the size and direction of a step cannot be predicted on the basis of the previous ones. Next, step sizes must be distributed according to the familiar bell-shaped curve (the scientific term being that they are "normally distributed"). Finally, and crucially, the paths that underlie the process must be continuous.

With the salient features determined, another question poses itself. Does a process that exhibits these four properties actually exist? At first blush this may seem like a superfluous question that only mathematicians who are in the habit of questioning even the most obvious could come up with. After all, Brown saw that the particles move around, and so did Svedberg, Perrin, and many others. So what's the big deal? Are not particles in a liquid or drunken sailors near a lamppost the best proofs of the movement's existence?

No, they aren't. It is conceivable that Wiener could have done too much of a good thing. Even though the four properties seem to accord well with the behavior of the particles and the sailors, he could have inadvertently stated properties that are mutually incompatible. He would have thus conceived of a "Wiener process" that cannot exist in nature. It is like, say, seeking a number that is odd, greater than 175, divisible by two, and smaller than 58. Since the requirements are incompatible, that number doesn't exist. So the fact that Brownian motion can be observed does not obviate the need to prove the existence of the Wiener process. Only if it actually exists can the Wiener process serve as a mathematical model of Brownian motion and the four properties can be used to simulate it.

Wiener proved that Brownian motion really does exist. Showing that there are processes whose observed steps are normally distributed was the easy part. In fact, one can simply design a process with normally distributed steps, so actually there is not much to prove. The difficult part was to show that the positions of a particle, or of a drunken sailor, at certain instances in time are discrete observations—samples if you will—of a *continuous* path.

Imagine the dance floor of a disco. Flashing strobe lights highlight the seemingly jerky movements of the dancers. Of course, even the coolest dancers do not disappear into thin air between the flashes. They move continuously through time and space, but their movements are hidden during the dark periods between flashes. In fact, there could be different dance steps that give rise to identical positions during the strobe flashes. This is the key point. Particles pushed around by the molecules in a liquid describe a continuous path, and the jerky motion we see under the microscope is just a sample of the entire movement. What Wiener proved was that even for the jerkiest observations, it is practically certain that continuous paths exist which give rise to them.

Wiener's proof is by no means trivial and is contained in his forty-page paper, "Differential Space," that appeared in the *Journal of Mathematics and Physics* in 1923. The essay's title refers to the fact that Wiener was not working with the values of a function, but with the difference between two consecutive values (i.e., with the particle's or drunkard's steps) from one observation to the next. The choice of terminology is somewhat ironic because the paths that give rise to the jerky observations are nowhere differentiable. This latter word is a technical term to describe a curve that is smooth. The path is never smooth, not even for very short stretches. It is kinked and jerky, and when one zooms in to see more clearly, the path is more kinked and jerky still.[6]

This is what threw off Svedberg when he tried to measure the speed of particles suspended in a liquid. Whenever he increased the magnification of his microscope in order to get better estimates, the particle's speed seemed to grow. The reason was that the path, with all its minute zigs and zags, seemed to become longer the closer he looked. Consequently the speed—the length of the path divided by time—seemed to become higher and higher. The bottom line is that paths that give rise to Brownian motion are nowhere sufficiently smooth to allow measurement of the particle's speed. In technical terms, the path is nowhere differentiable.

Wiener's work on Brownian motion was his first serious scientific contribution and it would occupy him on and off throughout his life. Some people consider it one of his most important achievements. In his autobiography Wiener recalls that "this study introduced me to the theory of probability. Moreover, it led me very directly to the periodogram and to the study of forms of harmonic analysis. . . . All these concepts have combined with the engineering preoccupations of a professor of the Massachusetts Institute of Technology to lead me to make both theoretical and practical advances in the theory of communication, and ultimately to found the discipline of cybernetics. . . . Thus, varied as my scientific interests seem to be, there has been a single thread connecting all of them from my first mature work."

Given Wiener's proof that the four properties are compatible and that paths which conform to them exist, these properties can serve to describe, and even define, Brownian motion. From the requirement that the steps be normally distributed it follows that forward and backward steps balance each other. Hence the average step size is zero but—as stated repeatedly and as can be proven mathematically—the distance covered during a certain time period is proportional, on average, to the square root of that time period. Thus the "Wiener process" is a good mathematical model of the physical "Brownian motion." Whenever a quantity is influenced by many independent perturbations, we see a Wiener process.

About fifteen years after Wiener's important paper, the French mathematician Paul Lévy—whose initially less than benevolent attitude toward Bachelier was described in an earlier chapter—developed the theoretical foundations of Brownian motion further.

Born in 1886 in Paris, Lévy was destined to become a mathematician. Both his father and grandfather were professors of the subject, and his father was even examiner at the École Polytechnique. Paul was consistently at the top of his class in mathematics but was no slouch in other subjects either. He received the Prix du Concours Général, awarded by the University of Paris to brilliant high school students every year since 1744, not only for mathematics but also for ancient Greek.

When he applied for entry to the leading schools of mathematics in France, Paul was ranked first in the entrance examination at École Normale Supérieure, and second at École Polytechnique. He nevertheless chose to attend the latter and graduated at the very top of his class two years later. After completing his one-year military service, compulsory for graduates of the Polytechnique, he attended École des Mines for three years. He later recounted that he would routinely skip classes at the engineering school in order to take the more advanced mathematics courses offered at the Sorbonne. In 1912 he obtained his doctorate and started teaching at the École des Mines the following year.

During World War I, he had to interrupt his career and serve in the French artillery corps. He mainly worked on antiaircraft defense—similar to the work that Wiener would be doing for the U.S. government during World War II. Returning to civilian life, he resumed teaching at Mines while also lecturing at Polytechnique.[7] In 1919 a professor at École Polytechnique, the prominent number theorist Georges Humbert, fell ill and asked Lévy to assume part of his teaching duties. When Humbert had to resign because of poor health in the following year, Lévy was appointed his successor with the rank of full professor. Apart from a break between 1943 and 1945, Lévy remained at the École Polytechnique until his retirement in 1959.

The break was due to the infamous laws of Vichy. After the defeat of the French army, the Polytechnique moved to the city of Lyon in the nonoccupied part of France and Lévy was able to continue teaching. In

1943, when all of France was occupied by the Germans, the Polytechnique moved back to Paris. But not Lévy. The Vichy government had voluntarily, without coercion by the Nazis, adopted the Statut des Juifs, a set of racial laws that excluded Jews from certain roles in society, such as teaching positions at universities. As a Jew, Lévy stayed behind, hiding somewhere in the south of France, and in this way survived the war.

Lévy was described as a kind and modest person. But there was another side to his character. For better or worse, he was quick to make up his mind about other mathematicians and their abilities, or the lack thereof. Bachelier was not the only colleague who suffered from his abrasive manner.

While he was substitute teaching for Humbert, Lévy, already a renowned professor of analysis, was asked by the dean of École Polytechnique to deliver a series of three lectures on "notions of Calculus of Probabilities and the role of the Gaussian Law [i.e., the normal distribution] in the theory of errors." This assignment induced Lévy to take a close look at probability, a subject that would occupy him for the rest of his life. Lévy turned the subject matter, which was then just a collection of computational problems, into a full-fledged branch of mathematics, and today he is rightly considered a founding father of probability theory. Of the 270 papers he published, 150 were on probability theory. Given his importance and fame, it is somewhat strange that he was elected to the Académie des Sciences only in 1964 when he was seventy-eight years old, one year after being elected to honorary membership in the London Mathematical Society.

Lévy studied the properties of Brownian paths for many years. In 1939 he proved the converse of Wiener's theorem: if the steps of a dynamic process are not normally distributed, then either consecutive steps are not independent of each other, or the path is not continuous. The strides of a drunken and increasingly tired sailor, for example, will get smaller as time goes by. Or a child running about in a playground will take smaller steps away from his mother as he gets farther away from her. Hence their walk will no longer correspond to the definition of Brownian paths. In 1948 he published the classic book *Processus stochastiques et mouvement Brownien* (Stochastic processes and Brownian motion).

Distributions other than the so-called normal, or Gaussian, were of special interest to Lévy. Today, a variant of the Gaussian distribution carries his name. It is the distribution of the steps of Brownian motion, interspersed every once in a while with a large displacement in a random direction. In distinction to the Brownian motion with its infinitesimally small steps, this process is called "Lévy flight." It occurs when a particle, a drunken sailor, or some other creature or object performs Brownian motion for a while, then carries out a long displacement, followed by Brownian motion at the new location, and so on.

Lévy flights are very common and have found many applications in the natural and social sciences. In contrast to Brownian motion, where the distances that particles travel is proportional, on average, to the square root of time, Lévy flights can cover wide areas. A particularly famous instance of Lévy flights was found in the foraging behavior of albatrosses in the Antarctic. On Bird Island in the South Atlantic, sensors were attached to the legs of albatrosses that detected and registered when they were wet or dry. In the first instance, the bird was in the water eating fish and crustaceans. When the sensor was dry, it was assumed that the bird was in the air, flying to new fishing grounds. Evaluation of the collected data showed that the flying patterns conformed to the Lévy distribution. Most foraging flights were short in the typical pattern of Brownian motion, but there were occasional long flights—up to ninety-nine uninterrupted hours—to new fishing grounds. Once there, the albatrosses' flying time again conformed to the Gaussian distribution, the characteristic required of Brownian motion.[8]

Brownian motion and Lévy flights belong to the class of random walks. Random walks can take place in one, two, three, or more dimensions, and they come in many flavors. There are self-avoiding random walks that never intersect, reflecting random walks that bounce off boundaries like billiard balls, loop-erased random walks that are left over after any loops of the original random walks have been erased, biased random walks, random walks with drift, and many other variations. In 2006 the prestigious Fields Medal, the highest scientific prize for mathematics, was awarded to the German-French mathematician Wendelin Werner for his work on random walks.

EIGHT

The Overlooked Thesis

W HEN SCIENTISTS LIKE EINSTEIN, PEARSON, WIENER, and Lévy advanced the understanding of random walks in the context of the natural sciences and engineering, they paid no attention to one potential area of application: the stock market. It was not their fault. People interested in economics at that time, mostly businessmen and lawyers, never brought their field to the notice of researchers trained in the use of mathematics. True, Jules Regnault's work had appeared in the 1860s, Henri Lefèvre's in the 1870s and 1880s. However, in spite of their tentative steps toward a more quantitative approach, the description of economic phenomena was still couched not in equations but in words. But with the work of accountants becoming increasingly prominent in economic circles, mathematics began to have some impact on economic research. Bachelier's thesis was the first attempt to breach the boundary between mathematics and economic theory.

Let us return to France, 1900. That year, both the World's Fair and the Olympic Games were held in Paris. To celebrate the turn of the century and the central position that Paris held in the intellectual world, more than a hundred academic conferences were also held in the French capital that year. At the International Congress of Mathematicians, the German mathematician David Hilbert proposed his famous list of twenty-three open problems that would largely set the agenda for mathematical research for the twentieth century.[1] At the Congrès International des Valeurs Mobilières (International Congress on Financial Securities), which was held in Paris just two months after Bachelier's thesis defense, several hundred financial economists, investment bankers, and corporate lawyers gathered at a meeting that was described by the prominent French statistician and economist Alfred Neymarck as a resounding success. In retrospect, both the congress and Bachelier's thesis, which coincided, can now be seen as the beginning of a general trend toward the mathematization of economics. Regnault had brought statistics to bear on the subject, Lefèvre geometry. Now Bachelier brought calculus into play.

One thing that strikes the reader of the thesis is the highly technical treatment, with dozens of complicated-looking equations and integrals scattered throughout the sixty-six pages of his work. In spite of this, it turns out that Bachelier was not rigorous and his arguments were often flimsy. Thirty years later, while admitting that the Frenchman's work had strongly influenced him, the eminent Russian probabilist Andrei Kolmogorov would note that "the Bachelierian treatment wholly lacks mathematical rigor." But then Bachelier's thesis supervisor, Henri Poincaré, the most important mathematician of the turn of the century,[2] was not known for exacting strictness, and this may explain why numerous inaccuracies and even errors in the thesis could make it past the examiners. However, again like his teacher, Bachelier's strong intuition unfailingly pointed him in the correct direction. Even today, Bachelier's work seems path-breaking. "There is sometimes a lack of rigor, but never a shortage of originality or sound intuition," exclaimed two mathematicians who had studied Bachelier's work intimately.[3]

Another striking element of Bachelier's work was his failure to reference the earlier work of others. At the time he was writing his thesis it was not customary, as it is today, to include exhaustive bibliographies. He probably knew Regnault's book and could have been inspired by it. And it is nearly certain that the manner in which he examines and explains options was borrowed from Lefèvre's graphical method.

Bachelier begins his work with a caveat. "The influences that determine the movements on the Bourse are uncountable. . . . [Hence] the determination of these movements is subject to an infinite number of factors and it is therefore impossible to hope for mathematical predictability." After all, contradictory expectations by buyers, who believe prices will rise, and sellers, who at the same time believe prices will fall, are just about equally divided. Hence, Bachelier cautions, "the calculation of probabilities can never be applied to the movements of stock prices, and the dynamic of the Bourse will never be an exact science."

But one thing that can be studied mathematically is the state of the market at any given time. Even though actual price quotations cannot be predicted, their probability distribution can be. Some price movements are more likely than others and the probability of price variations can be established. "La recherche d'une formule qui l'exprime . . . sera l'object de ce travail." (The search for a formula that expresses it . . . will be the object of his work.)

After these introductory remarks, the thesis delves into a series of definitions: *opérations fermes, opérations à primes, reports, rentes reportable, cours vrais, réponse des primes, écarts des primes,* and so on.[4] The *opération ferme* nowadays is known as a forward contract. It is equivalent in all aspects to the regular purchase of a security, except that payment for, and delivery of, the security are made at a specified date in the future, at a price specified in advance. If the price rises above the purchase price, the buyer profits from the deal; otherwise he loses. In fact, possible profits are unlimited, since the security's price could go sky high. On the other hand, the speculator could lose his entire investment since the share's value could fall to zero. Obviously the seller's profit or loss potential is the exact opposite of the buyer's potential.

The *opération à prime* is the transaction that is the subject of this book: the option. The buyer of an option, Bachelier explains, pays the seller a bit more than he would have paid for a spot or a forward transaction, but—in exchange—puts a cap on his losses. The amount he pays over and above the going price is called the premium (hence *à prime*). But, as pointed out previously, he need not take possession of and pay for the actual security if he does not want to. If its price drops in the meantime, he can walk away from the deal. In that case he pays the seller the premium, and his loss is limited to that amount. So while the buyer of a security on the spot or on the futures market can lose everything, the options buyer can, at most, lose the premium . . . however far the price of the security falls.

After explaining the bourse's operations, Bachelier suggests a geometrical method in order to understand options purchases. "Proposons-nous de représenter géométriquement un achat à prime" (Let us illustrate an options purchase geometrically), he writes, and presents a figure that illustrates when and how much an investor gains or loses from the purchase of an option.

There is something about the illustration that evokes a sense of déjà vu. The figure with the kinked line that specifies the profit possibilities, drawn in the *x*- and *y*-coordinate system, is identical to the representation of option purchases that Lefèvre suggested some thirty years earlier. Even though he does not attribute the geometrical technique to his predecessor, Bachelier almost certainly learned of it from Lefèvre's work. As we saw in Chapter 4, Lefèvre made quite a song and dance about his invention, and anybody involved in studying the options market after the mid-1870s must have been familiar with his graphs.

Bachelier then discusses variations in stock market prices. There are two kinds of uncertainties pertaining to the price of securities, he writes. One is the uncertainty about future events and how they will affect the market. These could be wars, bad harvests, bankruptcies, and so on. Such events are impossible to predict and cannot be treated mathematically, he cautions. Furthermore, the effect of such an occurrence on the market cannot always be predicted, since expectations are entirely personal. Some speculators believe that a certain event will result in a price increase; others are convinced that the price must decrease. And even if all speculators agree on the direction in which the price must move (e.g., a bad harvest will certainly lead to an increase in the price of wheat), the amount of the increase is uncertain.

Further reflection on this leads Bachelier to an important observation. Since there must be as many sellers as there are buyers for each transaction price, the speculators' expectations offset each other. (Given his background in physics, Bachelier could have pointed out that the stock market behaves in this respect according to Newton's Third Law, which says that to every action there is an equal and opposite reaction. But he didn't.) This leads Bachelier to the remarkable conclusion that "at a given moment the market believes neither in an increase nor in a decrease of the true price." This is a truly significant statement. It says that the most probable price in the future is the one that is in effect today. All information about the present state of affairs and all expectations about the future are contained in the current price. In modern financial theory this principle is known as the "efficient market hypothesis."

But even if actual prices cannot be predicted, and in spite of the fact that the market believes in neither a rise nor a fall of prices, deviations from the true price will occur. Bachelier maintains that one can make inferences about the likelihood and the magnitude of the price's divergence from the true price. The objective of his thesis was to investigate the probabilities that divergences of various magnitudes occur—to determine the distribution of the size of the movements.

Like Regnault forty years earlier, Bachelier maintains that since prices are just as likely to go up as they are to go down, and taking into account compound interest and the forward prices, *"l'espérance mathématique du spéculateur est nulle."* (The mathematical expectation [of the profit] of the speculator is zero.) This assertion, italicized in the original text for greater emphasis, is so important that Bachelier states it as an axiom, which he calls the "fundamental principle." It would serve as a basic postulate for the rest of the thesis.

In summary, the price that the market considers most likely in the future is exactly the price that is currently quoted on the forward market. If it were otherwise, the market would already quote a different price today. Hence the best estimate of tomorrow's price, or next week's, or next month's, is today's price. The inevitable, if depressing, conclusion is that if a speculator buys a security with the intention of reselling it, he may expect a zero profit.

Up to that point, Bachelier kept to the tradition of describing economic phenomena verbally. Only one equation appears on the first fourteen pages. But from page 15 on, and all the way to the end of the sixty-six-page thesis, there is hardly a paragraph that does not display at least one and sometimes several dense equations. And this is what distinguishes Bachelier's work from that of his forerunners: his expert use of mathematics allowed him to advance further than any of his predecessors.

Bachelier's investigations are guided by two principles: the axiom that the mathematical expectation of the speculator's profit is zero, and the assumption that fluctuations of the quotations around the true price are independent of the value of the stock's price. The latter says that a deviation of, say, 25 centimes from the true price is just as likely to occur if the true price is 20 francs as when it is 100.

Then the real work starts. What is the probability that the price of a stock moves a certain distance during a certain time period? Using the so-called principle of compounding probabilities, Bachelier answers this question by multiplying the probability that the price is x at the beginning of the time period, with the probability that it rises to z by the end of the time period. Next he considers all possible prices x that the share can assume, performs a few more mathematical manipulations, and is led to the result that he sought. Lo and behold, the conclusion is that the fluctuations of the stock's price around the true price are governed by the Gaussian distribution (i.e., the familiar bell curve).

The innovative use of probability theory to describe price movements on the bourse would lead Poincaré to write in his report on the thesis that "the manner in which Monsieur Bachelier extracts Gauss's law is very original." Having one's work described as "very original" by the master of originality is high praise indeed. As a by-product of his derivation, Bachelier obtained an equation that would be made famous thirty years later by Sydney Chapman, a British astronomer and geophysicist, and Nikolai Kolmogorov, a Russian pioneer of probability theory. In a chain of independent events, this equation relates the probability distribution of the variable at the end of the chain, to the probability distributions of the intermediate steps. In the landmark paper "Über die analytischen Methoden der Wahrscheinlichkeitsrechnung" ("On the Analytical Methods in Probability Theory), published in 1931 in the prestigious German journal *Mathematische Annalen*, Kolmogorov credits Bachelier's seminal thesis as the first work that uses probability theory to model dynamic processes.

Given that stock price movements accord with the bell-shaped Gaussian distribution, Bachelier next asks what displacement one may expect over time—what distance does the price cover, on average, during a certain time period? Lo and behold again! Bachelier finds that "l'espérence mathématique . . . est proportionnelle à la racine carrée du temps." (The mathematical expectation [of price movements] is proportional to the square root of time.) This is exactly what Einstein, Smoluchowski, and Langevin would learn five years later concerning the movement of molecules. Without knowing it, and before anybody else, Bachelier had discovered the process that governs Brownian motion. " $\sqrt{x}$ "

Not content with just one justification for the square root of time law, Bachelier gives a second one. It is combinatorial in nature, which is the manner in which Smoluchowski would be treating the problem. Bachelier assumes that the chances of upward and downward movements of prices are equal, which today is the standard assumption of random walks and Brownian motion. Like a good pupil, he goes through some of the mathematical manipulations quite explicitly—at one point giving no less than three different ways of evaluating an integral before mentioning that there are ready-made tables from which the values can be read off—while sailing through other calculations with nary a word of explanation. But then, a Ph.D. dissertation is as much about showing off one's erudition as it is about providing the world with new findings. It is therefore indispensable to demonstrate some proficiency in mathematical acrobatics.

This part of the thesis is fraught with inaccuracies and errors. For example, at one point Bachelier subdivides a time period into very small time slices but later lets the slices become large. In another place, he assumes a variable to be infinitesimally small, overlooking the fact that it is an integer and must therefore equal at least one. Surprisingly, neither Poincaré nor his co-referees, Paul Appell and Joseph Boussinesq, noticed these shortcomings. In spite of the slipups, Bachelier attained his main goal, which was to show, yet again, that the expected distance that a price moves during a time period is proportional to the square root of time.

Next Bachelier considers the probability that a price will be equal to, or greater than, a certain threshold. For this he takes a cue from the theory of heat that was developed by the French mathematician and physicist Joseph Fourier in the first quarter of the nineteenth century. Fourier, basing himself on Isaac Newton's theory of cooling, investigated how heat flows from hot spots to cold areas until a constant temperature prevails throughout. Bachelier devised a model that describes the "flow of probabilities" over time. Within a small time period, he reasoned, the value of a security can move to one of the neighboring prices. Hence with a probability of 50 percent each, the next period's price will come to lie at the notch above or below the current one. Looked at in another way, at the end of the period, every price will have inherited half of the neighboring prices' probabilities. Thus probabilities flow from prices that have a high probability toward prices with low probability and, with time, the probabilities diffuse across all prices in the same manner that heat diffuses in space. In the course of his deliberations, Bachelier obtained Fourier's famous heat equation, thereby establishing a fundamental connection between Brownian motion and the flow of heat.

Now Bachelier turns to the pricing of options, the main topic of his thesis. Recalling his fundamental principle, he restates, again in italics, that *"the mathematical expectation of the buyer of the option is zero."* As we will see, everything else follows from this simple-sounding statement.

Bachelier develops a mathematical expression that describes the relationship between the probability distribution, the forward price of a security, its exercise price, and the price of an option. The variables are pulled together in an equation that computes the investor's expected gain or loss. Since the latter must be zero by the fundamental principle, the right-hand side of the equation is zero. Turning to the special case where the exercise price equals the forward price, Bachelier solves for the option's price. The expression he obtains is, in his own words, one of the most important of his thesis. It shows how the value of an option depends on the time until the exercise date: "La valeur de la prime simple doit être proportionnelle à la racine carrée du temps." (The value of the simple option must be proportional to the square root of time.)

How exactly the option price depends on the square root of time, apart from the two just being proportional to each other, is described by an equation that also contains a term that Bachelier calls the "coefficient of instability." This coefficient indicates whether a security's price tends to hold steady or whether it moves all over the place. High instability indicates *un état d'inquiétude,* or a state of nervousness; low instability indicates a state of calm. The price of an option is then given in Bachelier's equation by the volatility, multiplied by the square root of time. Hence the more volatile the security, the higher the price of the option.

The discovery that an option's value depends on time and on the underlying security's volatility was an extraordinary achievement. It was the first time that the price of an option, as it should prevail on the market, was derived scientifically. Up until then, an option's intrinsic value was thought to be determined through trial and error by the market itself—based on demand and supply on the bourse.

An interesting question now poses itself. Buyers of options would like to know what the chances are that they will gain from their transactions. Obviously, if they are to make money from an options purchase, the eventual payoff must be higher than the cost of the option. So Bachelier computed the probability that the price of a security will rise until the exercise date by at least the amount that the speculator paid for the option. He knew that the formula must contain the number of days left until the exercise date, since he had just shown that the option's price depended on the square root of time. But he was in for a surprise: time dropped out of the equation. The answer to his question was a plain and simple 0.345. Yep, just a number, no strings attached. In other words, in 34.5 percent of the cases investors come out ahead, no matter how many days or months remain until the exercise date. The other side of the coin is that speculators lose money with a probability of 65.5 percent.

Note that this does not contradict Bachelier's fundamental principle, which said that the expected gain to the speculator is zero. The reason is that the probabilities must be multiplied with the amounts of gains and losses in order to arrive at the correct weighted average. For example, if one gains $10.00 with a probability of 34.5 percent, and loses $5.27 with a probability of 65.5 percent, the expected gain is zero: $10.00 \times 0.345 + 5.27 \times 0.655 = 0$.

Up to this point, Bachelier had limited his investigations to "at the money" options. But he then relaxed this restriction, examining options when forward price and exercise price are not identical, and presented a host of additional equations and results.

To scientists, the theoretical analyses and derivations are very appealing, but to practitioners who invest in the stock market the crucial question is whether they are correct. The way to ascertain the veracity of Bachelier's results is through a comparison of the results predicted by his equations with the prices that prevailed on the stock market. After all, his model includes some pretty hefty assumptions, like the lack of transaction costs, the fact that his model lets the price jump only to one of the neighboring prices, the absence of noise in the data, and the supposition that supply and demand on the bourse does in fact result in the correct price.

The following pages of the thesis are quite revolutionary. As I pointed out already, until Bachelier wrote his thesis, the theory of economics and finance was all words. Numbers were shunned and predictions were not subjected to econometric assessments. But Bachelier did not shy away from the ultimate test. He compared the results, as they derived from his model, with actual data from the bourse. This was an early example of a statistical confirmation of an economic model.

Reading this part of the thesis, Poincaré was skeptical at first. "One must not expect too exact a verification . . . [it] cannot be more than coarse," he wrote. Using data from the previous five years, Bachelier himself did not really expect a strict numerical match between calculated and observed variables. Rather, he hoped for a qualitative agreement between the two in order "to show that the market, unwittingly, obeys a law that rules it: the law of probability." By and large, the fit was reasonable: discrepancies between calculated and observed results usually did not exceed 10 percent. Thus Bachelier's claim that the bourse's unwitting use of the trial and error method of supply and demand is replicated by his model was corroborated. Let's quote Poincaré again: "The author of the thesis gives statistics which make the verification in a very satisfactory manner."

Next, Bachelier investigates the probability of a security attaining a certain price before some fixed date in the future. His analysis leads to a multiple integral of intimidating dimension; there are as many superimposed integral signs as there are days until the fixed date. At first the daunting equation seems inaccessible, but "the author resolves it with a short, simple and elegant argument" (Poincaré's words). The short, simple, and elegant argument would later become known as the "reflection principle" of Brownian motion.

The principle says that a particle, or the price of a security, continues to follow the path of Brownian motion even after it hits an insurmountable barrier. The same laws that govern a billiard ball that bounces back after it hits the cushion around the table, govern the particle's path, both before and after it is reflected off a barrier. Making use of this insight—with the price that is to be attained playing the role of the barrier—Bachelier reaches an interesting conclusion: the probability that a security attains or exceeds a certain price at a certain date is equal to half the probability that the price is attained or exceeded at any time before that date. Poincaré was impressed. He sounds nearly ebullient when extolling the "originality of this ingenious trick."

As a final encore before wrapping up, Bachelier dispenses information that could almost be considered practical advice. He computes that "there are two chances in three that one can, without loss, resell a forward contract against a simple option," or "there are four chances in ten that one can, without loss, resell a forward contract against an option with a premium of $2a$" (a being the price of the option), and the like.

And with this Bachelier concludes his thesis. In cannot be overemphasized how revolutionary it was. Probability theory, which forms the basis for his investigations of the stock market, had practically been ignored by mathematicians before the turn of the century. Apart from gamblers who could verify the accuracy of probability calculus through their own experience, nobody really believed that its use was appropriate in real-life applications. Even Poincaré, who taught the subject at École Polytechnique, lent it only lukewarm support. The theory did not gain recognition until the 1930s. But Bachelier was not cowed into silence even in 1900. "One sees that the current theory—with the help of probability calculus—resolves most of the problems to which the study of speculations leads." The book's final remark, already quoted above, puts it into a nutshell: "The market, unwittingly, obeys a law that rules it: the law of probability." Obviously Bachelier was far ahead of his time.

NINE

Another Pioneer

I N THE LATE 1990S, SWISS HISTORIAN WOLFGANG HAFNER was researching a book on financial economics when he tripped over an amazing find. As a journalist specializing in economics and sociology, Hafner made a name for himself by uncovering the role that financial derivatives (e.g., options) can play in the laundering of money.[1] While researching his book, Hafner asked his antiquarian bookseller to procure for him any early German language books on financial markets. The bookseller discovered and sent him a textbook published in 1920 by Friedrich Leitner from Berlin. Of more interest to Hafner than the book itself was a footnote referencing another book that "deals with the subject from a mathematical point of view." Mathematical economics before 1920? Hafner became curious and got hold of the cited volume.

The slim, eighty-page booklet was published in 1908 by an all but forgotten professor of "commercial and political arithmetic," what we now call actuarial science and insurance mathematics, at the Accademia di Commercio e Nautica (Commercial and Naval Academy) in the city of Trieste on the Adriatic coast. The author's name was Vincenz Bronzin and the work was titled *Theorie der Prämiengeschäfte* (Theory of options transactions). Trieste, today in Italy, was part of the Austro-Hungarian Empire until the end of World War I, which explains why Bronzin wrote his scholarly work in German. The city, which was home to the most important seaport of the Habsburg Empire, served as a hub for international trade and became a wealthy center for banks, trading houses, and insurance companies. Parallel with Trieste's ascent as a shipping and trading center, the need arose for commercial education. This explains the importance and reputation of the Accademia throughout the empire.

Leafing through the pages of math, Hafner came upon a model that he had not expected: an equation for the price of an option. Apparently the booklet he was holding in his hands was a precursor to the modern theory of options pricing. Amazed that he had never heard of either Bronzin or his work, Hafner sent an e-mail to Heinz Zimmermann, professor of finance at the University of Basel in Switzerland, and inquired whether he had ever heard of this obscure professor. Zimmermann had not and was at first doubtful. He, like most financial economists, knew of Bachelier's early contribution to the theory of finance. Suddenly another forgotten forerunner appears out of nowhere? Zimmermann was about to dismiss the information Hafner had sent him but the more he read, the more surprised he became. Soon his skepticism gave way to keen interest. After the rediscovery of Regnault, Lefèvre, and Bachelier, a new pioneer was found.

Two years younger than Bachelier, Bronzin was born the son of a sea captain in 1872, in the town of Rovigno, then in Austro-Hungary, today in Croatia. After graduating from high school, he traveled to the empire's capital, Vienna, to study engineering and mathematics. There he heard lectures by the famous physicist Ludwig Boltzmann, whose views on probability had a great influence on his work, as Poincaré's had on Bachelier. But his time in Vienna was not all spent studying. He was also an avid competitive fencer and a successful gambler. His youthful interest in games of chance would stand him in good stead later, when he turned his theoretical interest to speculation and options trading.

Bronzin returned to Trieste and taught high school mathematics for a year and a half before being named professor of commercial and political arithmetic at the Commerce and Naval Academy. Professors were generally expected to publish scholarly material for educational purposes, and Bronzin has four known publications: a four-page article from 1904 in a journal for commercial education, two books, and a paper about the calculation of Easter in the Gregorian calendar. His first book, *Lehrbuch der politischen Arithmetik* (Textbook on political arithmetic), published in 1906, was approved by the ministry of eduction for use at commercial schools and academies throughout the empire. His second book is the one Hafner found. It was published in 1908 by the Franz-Deuticke Verlag in Vienna, which had also published Sigmund Freud's *Traumdeutung* (Interpretation of dreams).

Two years after publication the booklet was reviewed in the *Monatshefte für Mathematik und Physik,* an important monthly journal of mathematics and physics. Unfortunately the reviewer, the journal's founder Gustav Ritter von Escherich, professor of mathematics at the University of Vienna, attached little importance to Bronzin's book. In a nine-line review he commented that "it can hardly be assumed that the results will attain a particularly practical value." Since Escherich's reviews set the agenda in large part among mathematicians, it is little wonder that his prophecy was somewhat self-fulfilling.

In later years, Bronzin apparently never mentioned this work, and maybe even tried to ignore it. This may seem strange, but Hafner and Zimmermann remind us that trading in options, like gambling, was frowned on at the turn of the nineteenth century. Speculating in derivatives was blamed for excessive market movements and was considered socially harmful. As noted in the previous chapter, it was still very unusual to apply probability theory to transactions on the stock exchange. That, together with the bad review, may explain why the respected professor of political arithmetic did not consider it opportune to remind the public of his work. If Bronzin had wanted to make the public forget his work on options pricing, he succeeded because forgotten it remained . . . at least until it was unearthed by Hafner and Zimmermann nearly a century later.

Contemporaneous and subsequent references to Bronzin's work were few. Theoretical mathematicians may not have wanted to get their hands

dirty with something as mundane as the stock market, while insurance actuaries and accountants lacked the necessary tools to understand this newfangled model. And in the sixteen years that followed publication of the
booklet, economists cited Bronzin only three times before his book sank
into complete oblivion. In 1911 the Czech mathematician Gustav Flusser
mentioned Bronzin's work in a thirty-page paper on the pricing of options,
without adding anything substantial. In 1920 the German economist
Friedrich Leitner, who taught at the Handelshochschule in Berlin, a business school that later became part of Humboldt University, cited Bronzin's
work in his textbook on banking techniques. That was the book the antiquarian bookseller had found for Hafner. Finally, in 1924, Bronzin's booklet
was also mentioned in a textbook by Karl Meithner, a professor at the
Hochschule für Welthandel in Vienna (Academy for International Commerce).[2] Bronzin's findings fared no better among practitioners. Traders,
brokers, and speculators lacked the analytical training required to understand the practical relevance of his work. But when Hafner and Zimmermann analyzed the booklet's content, they recognized its true value.

At the turn of the twentieth century, not everybody believed in the existence of atoms and molecules, as we saw in Chapter 6. But those who did,
in particular those who believed that gases and liquids were made up of
them, also believed that Isaac Newton's laws of mechanics must apply to
these little balls. But there was a problem. Newton's Laws are reversible: if
a film of two objects knocking into each other is run backward, the resulting
scene is just as reasonable as the original one. Gases and fluids, on the other
hand, do not allow reversibility. Pour a drop of ink into a glass of water and
the ink diffuses. But the diffused ink would never coalesce back to a drop.
Heat a cold room and after a while it becomes uniformly warm throughout.
But the room would never cool down spontaneously with all the heat con-
centrating in the corner where the stove stands. Spray perfume on yourself
and your entire self smells nice. But the perfume particles would never
move back to the bottle's nozzle by themselves.

So, do Newton's laws fail to hold for gases and liquids? Boltzmann,
whose lectures so inspired Bronzin, thought not—and proved it. He

showed that Newton's laws hold in the small—on the level of individual atoms and molecules—but statistics take over when many, many, many atoms fly in all directions. In such a case, the so-called entropy of any open system must always increase, which is a fancy way of saying that disorder in the system always increases. Thus while each atom or molecule, by itself, obeys Newton's laws, the collection of billions and trillions of particles tends toward a state of maximum disorder that cannot be reversed. *Voilà!* Both Newton's laws and everyday observations of phenomena in gases and liquids were vindicated. Statistical mechanics was born. Most importantly, this is how probability enters the picture. For the first time the deterministic picture of everything physical was superseded by a statistical viewpoint. The most probable state in which a system can find itself is the one with the lowest entropy—the state with the highest disorder.

Not everybody bought it and emotions ran high. Resistance to the use of probability theory in physical systems was ubiquitous. Even Albert Einstein weighed in with the declaration that "God does not play dice." But Bronzin, like Bachelier before him, was convinced of the newfangled theories. And like Bachelier, he would apply the statistical approach to the behavior of stock markets. Both men realized that the market prices of securities contain all information available to speculators, that they were inherently unpredictable, and that probability theory was needed to price options and other derivatives.

Bronzin, like Lefèvre and Bachelier before him, also recognized that expected profits of both buyers and sellers must be zero. To derive the prices of options, he then proceeded to develop the same formula as Bachelier, an average of the profits to the speculator, weighted by the probabilities of different prices of the underlying security.[3] Of course, the expression, which consists of two integrals, depends on the probability distribution of the security's price.

To illustrate his result, Bronzin derives the pricing formulas for six different price distributions, four of which have no practical relevance and were used as examples only. But two of them, the so-called binomial distribution in which stock prices can go up or down by one tick in each

period, and the bell-shaped Gaussian distribution do resemble real-life behavior of the stock market. For the Gaussian distribution, the second integral is unsolvable. But as already pointed out by Bachelier, tables exist in
which the numerical result can be looked up. The result is a formula that—
up to some rewriting and reinterpretation of variables—is very similar to
the one that the heroes of this book's later chapters would find in the 1970s.

A lingering question remains. Did Bronzin know about Bachelier's earlier work? After all, Bachelier's Ph.D. dissertation was published in the *Annales Scientifiques de l'École Normale Supérieure* in 1900 and the gist of its
content could easily have found its way to Trieste. At least one Italian economist, Alfonso De Pietri-Tonelli from the Istituto Superiore di Scienze Eco-
nomiche (Institute of Economic Science) in Venice, quoted Bachelier
repeatedly in his book *La Speculazione di Borsa* (Speculation on the stock
exchange), albeit in 1919, eleven years after Bronzin had published his
booklet. Bronzin would certainly have been qualified to understand the
thesis. So could he have usurped Bachelier's findings without citing them?

Honi soit qui mal y pense (shame on him who thinks ill)—Zimmermann and Hafner sum up their thoughts on the question.[4] Even though
suspicion cannot be dispelled with certainty, it seems unlikely that the professor would stoop to plagiarizing someone else's work. Furthermore,
Bronzin's approach was quite different from Bachelier's. While falling short
of his predecessor's work in some respects (e.g., it lacked the element of
time) it also contained novel elements, such as the notion of arbitrage or
riskless profits, to which we will return in later chapters.

TEN

Measuring the Immeasurable

F OR THIRTY YEARS BACHELIER'S THESIS LANGUISHED UNNO-
ticed on bookshelves. As recounted in Chapter 5, Paul Lévy, one
of the few mathematicians who was aware of the book, erroneously thought he had discovered a mistake and dismissed it. The author's lack of a first-class education was another reason. But the absence
of academic recognition was not only due to personal shortcomings in
Bachelier and his colleagues. Probability theory and statistics still had to
overcome their disreputable standing. There was no room in science for
a discipline that used concepts such as luck, chance, and divine interven-
tion. Surrounded by an aura of seediness, probability theory was considered, at best, a speculative explanation for thermodynamics, at worst a
tool for card players and gamblers.[1]

A further cause for the general disdain in which probability theory was
held by serious scientists was the lack of a mathematically rigorous basis:

the newfangled theory was not built on solid foundations. Admittedly, the lack of sound theoretical underpinnings had not kept men of Boltzmann's and Einstein's stature from making use of this novel tool but then again, they were physicists, not mathematicians. Like engineers who go by the motto "if it ain't broke, don't fix it," physicists subscribe to the maxim "if it works, it's good enough."

But for a theory to be valid, mathematicians require the existence of a few simple, self-evident axioms, like the ones that Euclid developed for geometry, from which everything else follows. They must all be essential, not lead to any contradictions, and allow the deduction of everything the theory has to offer. At the turn of the nineteenth century, probability lacked solid foundations. Whatever was known was ad hoc; there was no axiomatic system and hence nothing that could, strictly speaking, be considered a theory. This was the reason for mathematicians' unease in regard to it. As we will see below, it was only in the 1930s that the Russian Andrei Nikolaevich Kolmogorov provided a system of axioms after some important preparatory work by his contemporary, Henri Lebesgue in France.

The problem was not something created by zealous purists. For example, the expected value of an action is a fundamental notion in probability theory. Participants want to know, for example, what profits they may expect when playing the stock market or what losses they face when gambling at a casino. The expected value is computed by multiplying the outcomes with the probabilities of their occurrence and summing the results. For example, the expected number of pips that appear when dice are thrown is ⅙ times one, plus ⅙ times two, plus ⅙ times three, plus ⅙ times four, plus ⅙ times five, plus ⅙ times six. This gives 3.5, which is the expected number of pips at the throw of a dice. The fact that the "expected" number of pips can never actually occur is only one of the incongruencies of probability theory that so bothered mathematicians at the time.[2] But at least the computational method seems straightforward enough, doesn't it? Well, it is not quite as simple as that. What if the number of potential outcomes is infinitely large and the probabilities are infinitely small?

To illustrate, imagine a game in which a number between zero and one is picked at random. If the number happens to be rational, like ⅝ or ⁷⁸⁹⁄₃₀₅₆ (= 0.25818063), the player gets a dollar. If the number picked is irrational, like 0. 25818063 . . . (with a never ending string of decimals following), he gets nothing. What is the expected value of this game?

There are infinitely many rational numbers and infinitely many irrational ones. Is the expected value, therefore, half a dollar? Certainly not! In the late nineteenth century the German mathematician Georg Cantor showed that there are many more irrational numbers than there are rational ones. The quantity of irrational numbers is of a higher sort of infinity than the quantity of rational ones. Hence the expected value of the game is certainly not 50 cents. So how does one multiply infinitely many ones and even more zeros with infinitely small probabilities? No wonder mathematicians were wary. Questions such as these made them unwilling to use the ostracized theory.

It took the skill of Henri Lebesgue to show the way out of this predicament. The son of a printer, Lebesgue was born in 1875 in the town of Beauvais, nowadays about an hour's drive on the motorway north of Paris. After preparation at two lycées, Lebesgue passed the entrance examination for the École Normale Supérieure in 1894 and upon completion of his studies could consider himself a member of France's mathematical establishment— in pointed contrast to Bachelier. Teaching appointments at universities followed, first in the French province then in Paris. Lebesgue became a highly respected mathematician, was elected to several learned societies throughout Europe, was bestowed honorary doctorates, and was awarded many prizes. He died in Paris in 1941.

Lebesgue aimed to rehabilitate the theory of integration, not put probability theory on a sound footing. Integration, formulated in the late seventeenth century independently by Isaac Newton and Gottfried Leibniz as a part of calculus, had been used since Archimedes' time, and even earlier in ancient Egypt, to compute the areas and volumes of geometrical objects. Computations were done by slicing the objects, be they two-dimensional figures like circles or three-dimensional bodies like pyramids, into smaller and smaller pieces and computing the areas and volumes of these pieces. Toward the middle of the nineteenth century, Bernhard Riemann of the University of Göttingen in Germany formalized the method to make it palatable to the mathematics community.

But there was a problem: Riemann's rigorous definition was valid only for so-called continuous functions—functions whose graphical representations are smooth, without any breaks or kinks. Lebesgue would change that. On April 29, 1901, a year before completing his doctorate, the twenty-five-year-old student presented a paper to the Académie des Sciences in Paris, "Sur une generalisation de l'intégrale définie" (On a generalization of the definite integral). Later published in the society's proceedings, it would usher in a new age of integral calculus, allowing subsequent mathematicians to extend integration to noncontinuous functions.

In this important paper Lebesgue established the notion of a measure. It is an abstract concept that can be imagined for a straight or curvy line simply as the line's length, for a geometrical figure as its area, and for a three-dimensional body as its volume. Lebesgue's measure, usually postulated to lie between zero and one (i.e., between zero and 100 percent) must satisfy some reasonable-sounding properties. For example, the measure of an object—a line, a surface, a body—must always be positive, the measure of a partial object must be smaller than the measure of the entire object, and the measure of the combination of several objects must be the sum of the individual measures.

So far, not much new, but appearances are deceiving. Lebesgue's precise definition of a measure vastly increased the practicability of calculus. Fifteen years later, the *Bulletin of the American Mathematical Society* was still gushing: "This new integral of Lebesgue is proving itself a wonderful tool. I might compare it to a modern Krupp gun, so easily does it penetrate barriers which before were impregnable." This was in 1916, in the middle of World War I, when comparisons with war materiel, especially German made, were the height of a compliment. The reviewer's comparison was apt: with Lebesgue's definition, calculus could penetrate into new areas of mathematics, which had hitherto been closed to it. And in 2001, the

Académie des Sciences decided to commemorate the centenary of the publication of Lebesgue's note by reprinting it, together with comments, in its *Comptes Rendus*. The measure's advantage lies in its very abstractness, which not only makes it applicable to continuous lines but extends its usefulness to sets of points that cannot readily be identified as lines.

By the way, what is the expected value of the game where the player gets a dollar if a rational number is picked and nothing if an irrational number is picked? Well, according to Georg Cantor and Henri Lebesgue, the rational numbers have a measure of zero, while the measure of the irrational numbers is one. This is so because rational numbers are isolated points, surrounded by irrational numbers. In other words, within any neighborhood around a rational point, and be it ever so small, there are infinitely many irrational points.[3] True, there exist infinitely many rationals, but they are all isolated, like islands in a sea of irrationals. To illustrate, the combined surface of all the islands would still be vanishingly small compared to the vast ocean. And so it is with rational numbers: a set of isolated points, even if there are infinitely many of them, has a measure of zero. So don't ever pay anything to partake in such a game. Its expected value is zero and you will end up getting nothing in return.

The establishment of Lebesgue's measure paved probability theory's way toward respectability. The task of breaking the ground fell on Andrei Nikolaevich Kolmogorov. Throughout his long and productive life, Kolmogorov advanced, indeed spawned, many fields of mathematics. He was born in 1903 in the city of Tambov, 400 kilometers southeast of Moscow, where his mother, a member of Russia's nobility, was staying during a journey from the Crimea. Tragically, she died in childbirth and Andrei Nikolaevich was brought up by his aunt.[4] Kolmogorov attended Moscow University to study mathematics but also took courses in metallurgy and history. Indeed, his first academic treatise was on landholding in Novgorod during the fifteenth and sixteenth centuries and he thought of publishing it. His professor was not very impressed, however, and scolded him for only providing one proof for his thesis. Aware of his student's other interest he remarked that while a single proof may suffice in mathematics, historians prefer at least ten proofs. As disappointing as the professor's reaction might have been, the episode was a fortunate one for mathematics, since it resolved any doubts the budding scientist may have had about his choice of career. Henceforth Kolmogorov devoted himself wholly to mathematics and physics, where one proof suffices.

He made his mark early on. As a freshman, while still wavering between history and math, he managed to disprove a hypothetical assertion that one of the lecturers had put forth, and immediately attracted the attention of professors and older students. In 1922, the nineteen-year-old student worked as a train conductor to make ends meet. Inside the cabin of a moving train, he came up with a surprising idea. A century earlier, the physicist and mathematician Joseph Fourier (the one of the heat equation) had devised a tool, thereafter called Fourier series, to approximate any mathematical function as long as it fulfills certain conditions. Now, alone in his train cabin, Kolmogorov thought up an example of a function that was, in fact, approximated nearly nowhere by its Fourier series. The result would make him an instant mathematical celebrity.

By the time he finished his graduate studies in 1928, he had eighteen mathematical papers to his credit. And there was no stopping him. Kolmogorov became one of the most eminent mathematicians of the twentieth century with many honors, prizes, memberships in learned societies, and honorary doctorates bestowed on him. His collected works fill three volumes with seminal papers in many fields: integration theory, functional analysis, topology, trigonometric series, the theory of sets, ergodic theory, probability, statistics, turbulence, celestial mechanics, classical mechanics, geophysics, information theory and cybernetics, control theory, and approximation theory. During World War II, he even dealt with ballistics and toward the end of his career he devoted himself to mathematical logic and the teaching of mathematics in secondary schools. So wide-ranging were his interests and so spectacular were his accomplishments that when an American professor visited Moscow in the 1960s, he said that he had come to check whether Andrei Nikolaevich Kolmogorov was one individual or a whole institution.[5]

One question of particular concern to Kolmogorov—as indeed to all of mankind—was whether the solar system is stable. Since Johannes Kepler computed the orbits of celestial bodies and Isaac Newton discovered the gravitational force, we do not fear that Earth, Mars, or Venus could deviate from their elliptical orbits around the sun one day and take a different track. We consider it self-evident that our planetary system will remain stable. But are we right in doing so? Could a visiting comet create a disturbance in our gravitational fields that would throw planets off kilter?

When Kepler, imperial mathematician to Emperor Rudolf II in Prague, computed the orbit of Mars, he failed to notice that the celestial movement did not correspond precisely to an ellipse. Even though the observational data that he had obtained (more exactly, stolen) from his predecessor[6] were the most exact at the time, they hid minute deviations from an ellipse, and in actual fact, the orbits were only quasi-periodic. The reason for the deviation from a perfect ellipse is that a planet's movement is determined not only by the sun's gravitation but also by the gravitational pull of all the other heavenly bodies. Newton's path-breaking new physics, however, related to the interaction of only two bodies. And as so often in life, adding a third body—a *ménage à trois* so to speak—makes a system vastly more complicated.

Mathematicians tried to compute the orbits of three or more bodies whose gravitation influences each other and soon ran into difficulties. They suspected that the so-called three-body problem could not be solved. Hoping to get young colleagues interested in the problem, the Swedish mathematician Gösta Mittag-Leffler suggested it to King Oskar II of Sweden and Norway in 1885 as a prize question in honor of the king's upcoming sixtieth birthday.

Two years later, twelve papers had been submitted. None of them proposed a solution, but one of them stood head and shoulders above the others—the one submitted by the French mathematician Henri Poincaré. He had at least been able to express the orbits of three bodies as mathematical series (i.e., as sums of infinitely many terms). But it was unclear whether the series converge to a finite result or whether the sums explode toward infinity. (This was just like the diverging Fourier series that had founded Kolmogorov's fame.) If the series do not converge, the possibility exists that Mercury, Venus, Earth, Mars, Jupiter, Saturn, Uranus, Neptune, and Pluto will someday go off course and vanish into the vastness of the infinite universe.

The reason for the difficulty in ascertaining whether or not the series converge is that the terms contain ever smaller denominators. Thus, even though the numerators become smaller, it is not clear what happens to the terms themselves. As a result, it is not possible to compute whether our solar system is stable. The "small divisor problem" as it was henceforth known, would keep generations of mathematicians busy.[7]

The value of Poincaré's paper cannot be overestimated, since it established the theory of dynamic systems and introduced so-called chaotic motion. Even though the examination of chaotic phenomena had to wait for the help of computers in the last quarter of the twentieth century, Poincaré had already realized that minimal disturbances could lead to large effects in a system (the so-called butterfly effect). So, is the solar system not as stable as we thought?

The question of whether gravitation can keep a system of planets like our solar system together, or whether orbits can diverge toward infinity in response to small disturbances, remained unanswered until the Congress of Mathematicians in 1954 in Amsterdam. The hero of the last day of this conference, the largest of its kind, was none other than Andrei Nikolaevich Kolmogorov. In a keynote address to the thousands of mathematicians who had come from all over the world he calmed most fears. He talked about the three-body problem, asking what happens to the periodic orbits of bodies when small disturbances get in the way. His answer was that disturbed orbits could become quasi-periodic but would not explode.

Unfortunately, there was a problem. When an editor of *Mathematical Reviews* asked the German mathematician Jürgen Moser to prepare a summary of Kolmogorov's paper, the young man, who had just recently earned his doctorate at Göttingen, was taken aback. The proof of Kolmogorov's main contention seemed to be incomplete. Even today experts are of different minds as to whether Kolmogorov proved that the many-body system is stable or quasi-stable, or if he only provided the basic ideas, with the details of the proof provided by his successors. In any case, Moser labored on the problem for seven years before he was able to close the supposed gap. He solved the problem of the small divisors by proving that the denominators in Poincaré's series became smaller at a faster rate than the numerators, and thus the sum converges. Moser had found the missing piece of the puzzle.

At the same time in Moscow, one of Kolmogorov's students also fiddled with the problem. Vladimir Igorevich Arnold attacked the many-body problem from a different perspective. While Moser analyzed disturbances that were sufficiently smooth, Arnold investigated disturbances that could be expressed as series. The two arrived at the same conclusion as Kolmogorov: orbits could degenerate to quasi-stability but would not explode. In honor of the three mathematicians, the theory that grew out of their work is named KAM theory after their initials.

With this, the question of the solar system's stability could finally be answered: planets will not deviate from their usual orbits for astronomically long times. Hence humankind does not have to worry about the stability of our solar system, at least for the next few billion years or so. Actually, this is saying a bit too much. KAM theory refers to the three-body problem. But the solar system consists of the sun and eight planets, not to speak of the gravitational pull of bodies from outside our solar system. So a slight unease may befall the reader nevertheless: maybe, just maybe the solar system could explode anyway in a few billion years? Watch out for new developments in KAM theory.

Kolmogorov was an outdoor man who loved swimming, walking, boating, and skiing. His colleagues and students were expected to participate in the sports activities, which also served as opportunities to discuss mathematical matters. A fruitful collaboration with a colleague, the mathematician Pavel Sergeevich Alexandrov, started with a 21-day, 1,300-kilometer boat journey along the Volga during the summer of 1929. Later the two friends bought a house together on the bank of a river that became a center of mathematical activity for half a century. It has since been transformed into the Kolmogorov-Alexandrov memorial.

The bond between Kolmogorov and Alexandrov, which lasted until the latter's death in 1982, went far beyond the purely mathematical. Apart from sharing a house, they also shared hotel rooms at scientific conferences. It was generally known that they had a homosexual relationship, although they never acknowledged their liaison since homosexuality was a criminal offense in the Soviet Union. The constant threat of exposure and possible persecution by the secret police forced the two to publicly denounce Alexander Solzhenitsyn, for example.[8]

Kolmogorov also contributed to the Soviet war effort. In September 1941, three months after Adolf Hitler broke the Molotov-Ribbentrop pact with Joseph Stalin and launched an attack against the Soviet Union, Kolmogorov wrote a paper on the theory of firing dispersion, "Estimation of the Center and Spread of Dispersion for a Bounded Sample." In another paper, "On the Number of Hits Among Several Shots and General Principles of Characterizing Efficiency of Firing Strategies," Kolmogorov defines an efficiency criterion of a firing strategy. The papers were written in response to a request from the army to "give his conclusion on the subject of discrepancies among the existing methods to estimate precision from experimental data." Strangely enough, the papers were published during the war for all interested parties to see, and so the enemy could also read them. In 1948 the work of Kolmogorov and collaborators on firing dispersion was translated by the RAND Corporation under the title "Collection of Papers on the Theory of Artillery Fire." It is still regarded as important in U.S. military education.

For our purposes, Kolmogorov's most important achievement was a monograph on probability theory that he published in 1933. This pathbreaking treatise answered the challenge, alluded to in Chapter 8, that the German mathematician David Hilbert from the University of Göttingen, then—together with Henri Poincaré—the undisputed mathematical leader of the world, had posed in Paris in 1900, three years before Kolmogorov was born.

Hilbert had been asked by the organizers of the Mathematical Congress in Paris to formulate the most important problems that faced the mathematical community in the coming century, and Hilbert identified twenty-three. For the next half century, and even beyond, his program dictated mathematical activity around the world. Today, most problems on Hilbert's list have been solved. The first one to go was problem 3—given any two polyhedra of equal volume, is it always possible to cut the first into finitely many polyhedral pieces that can be reassembled to yield the second? Hilbert conjectured that this is not always possible. The question was motivated by the formula for the volume of pyramids. It had been known at least since the time of Euclid but its derivation relied on infinitesimal methods that would become calculus. Hilbert wanted to know whether elementary methods were sufficient to derive the formula. In the same year that he presented the problems in Paris, his student Max Dehn proved that the answer in general is no by producing a counterexample. He used algebraic methods to prove the impossibility.

The most recent advance was the answer to the third part of Hilbert's problem 18, which inquires about the shape of space. It asks about the densest packing of spheres, otherwise known as Kepler's Conjecture. The proof provided in 1998 by American mathematician Tom Hales—which showed that the densest way to pack spheres is to stack them in pyramids as fruit vendors stack oranges and apples—remains controversial, however, because it was computer assisted.[9] Problem 13, by the way, a question about seventh-degree equations, was solved by our very own Andrei Nikolaevich Kolmogorov, together with his student Vladimir Igorevich Arnold, in 1956 and 1957. Some problems were resolved differently than Hilbert would have thought or hoped. For example, problem 2 asks for a proof that the axioms of arithmetic are consistent. But it was shown by the logician Kurt Gödel that there can be no such proof. Or, referring to Diophantine equations with any number of unknown quantities and with rational integral numerical coefficients, problem 10 demands "to devise a process according to which it can be determined in a finite number of operations whether the equation is solvable in rational integers." Again, in 1970 the twenty-three-year-old Russian mathematician Yuri Matiyasevich showed that no such algorithm can exist. More than a century later, problem 8, the famous Riemann Conjecture, and 12, another problem in number theory, remain clearly unsolved.

A few of Hilbert's problems are too vague to be completely resolved. Among them is problem 6, which provided the launching pad for Kolmogorov's famous 1933 monograph "Grundbegriffe der Wahrscheinlichkeitsrechnung" (Foundations of probability theory).

Hilbert was inspired to formulate problem 6 by the beauty of Euclidean geometry. All there is to know about lines, triangles, and other geometrical figures can be derived from five simple, self-evident axioms. Hilbert wanted "to treat in the same manner, by means of axioms, those physical sciences in which already today mathematics plays an important part; these are, first of all, the probability calculus and mechanics." The part about mechanics is presumably too broad to ever be satisfactorily answered, but Kolmogorov in effect met Hilbert's challenge concerning probability, thus at long last gaining respectability for the controversial theory.

"The purpose of this monograph is to give an axiomatic foundation for the theory of probability," Komogorov wrote in the preface of his booklet and then continued, speaking of himself in the third person, "the author set himself the task of putting in their natural place among the general notions of modern mathematics, the basic concepts of probability theory—concepts which until recently were considered to be quite peculiar." He continues with a tip of the hat toward Henri Lebesgue. "The task would have been a rather hopeless one before the introduction of Lebesgue's theories of measure. However, after Lebesgue's publication of his investigations, the analogies between measure of a set and probability of an event . . . became apparent."

In Chapter 1, Kolmogorov specifies five axioms that would henceforth form the basis on which probability theory is founded. Everything else follows from them. As expected, they sound very reasonable; in fact, they are self-evident. For example, axiom 3 states that each possible event, or set of events, is assigned a number between, and including, zero and one. This number is called the event's probability and is its Lebesgue measure.[10] Axiom 4 says that the probability that *some* event will happen, from among all events that could happen, is one. The axiom guarantees that the "events space" includes all eventualities. The events space of coin tosses, for example, includes the coin landing heads, with a probability of one-half, the coin landing tails, also with a probability of one-half, and the coin landing on its edge, with a probability of zero. Kolmogorov stresses that a zero probability does not necessarily mean that the event is impossible, only that it is extremely unlikely. On one throw, it just won't happen. But if you toss the coin millions of times, it is not totally inconceivable that it will sometimes land on its edge. Yes, this does sound a bit confusing, and that is why probability theory was considered so peculiar.

Axiom 5 states that the combined probability of either of two mutually exclusive sets of events occurring is the sum of the individual probabilities. To illustrate, if the probability of rain is 25 percent and the probability of snow is 10 percent, then, according to axiom 5, the probability of either rain or snow falling is 35 percent. While axiom 5 refers to just two sets of events, it is easily extendable to three and more. For example, after having combined the events "rain or snow" into one set with a combined probability of 35 percent, another set could consist of the event "hail" with a probability of 5 percent. Then, by axiom 5, these sets could be combined to "rain or snow or hail" with a probability of 40 percent. But it is not permissible to continue like that for infinitely many events. In that case another axiom is needed, and in Chapter 2, Kolmogorov extends axiom 5 to infinitely many sets of events by positing axiom 6, which is known as the "axiom of continuity."

Let's verify the game described previously, where you get a dollar if a number, picked at random between zero and one, turns out to be rational. Since the probability of hitting a rational number is zero, and the probability of picking any number is one, it follows from the axioms that you are sure to pick an irrational number. But, like the coin toss, it is not impossible, in principle, that a rational number be picked. If you play the game a trillion times, you could just happen to be lucky.

All kinds of facts and theorems can be derived from the six axioms, for example, rules for the calculation of probabilities. To illustrate, the probability that an event will not occur is one, minus the probability that it will.

Or the probability that A or B will happen is the probability that A will happen, plus the probability that B will happen, minus the probability that both A and B will happen. (The latter probability is deducted in order to avoid double-counting.)

So Kolmogorov devised a nice formalistic model of something. But what? How do these axioms relate to phenomena that were usually associated with probability? In Section 2 of Chapter 1, "Das Verhältnis zur Erfahrungswelt" (The relation to the real world), Kolmogorov addressed the issue. In two pages he established how the axioms connect with the realm of observations and experimental data.

The question was whether probability describes a real, physical tendency of something to occur, or whether it is just a measure of how strongly one believes it will occur. One way to interpret an event's probability is by defining it as the frequency with which it occurs. Another interpretation of probability is inherent in degree of belief statements, as in, "based on the evidence presented to the court, there's a good chance that the accused is guilty" or "given what scientists know, dinosaurs probably became extinct after a meteor hit the earth."

Kolmogorov subscribed to the former interpretation of probability, thus adhering to the so-called frequentist position. It says that if an experiment is repeated many times, the ratio of successes (m) divided by the number of trials (n) (i.e., m/n) is understood as the event's probability. With this, Kolmogorov formalized a position that had already been taken by the Swiss mathematician Jacob Bernoulli two hundred years earlier. In his famous treatise *Ars Conjectandi* (The art of conjecturing) of 1713, Bernoulli had shown that in a sufficiently long series of trials, it is very likely that the frequency with which an event occurs is close to its probability.

Kolmogorov added a caveat, however: the definition is strictly correct only as n grows very large (i.e., as the number of trials goes toward infinity). Thus if the probability of an event is very small, it is practically certain that the event will not take place on one realization. But even if the series of trials is very long, the result may be off. When tossing coins, for example, it cannot be expected that the frequency of successes approaches 0.5000000 every time, even for long series.

The monograph represented a watershed. By putting probability calculus on an axiomatic footing, the booklet initiated a new era for probability and its methods. The stage was set for further advances. Building on the Russian mathematician's rigorous foundations, successors would be able to broaden the understanding of random events. From 1933 onward, physicists, gamblers, and stock market players could finally use probability calculus with confidence. It was no longer considered peculiar; it had been elevated to a theory.

When Paul Lévy first read Kolmogorov's "Grundbegriffe," he nearly kicked himself. Eight years earlier he had written his book *Calcul des probabilités* (Probability calculus). Lebesgue's measure theory had been around for a quarter century and his own book contained most of the machinery that would have been necessary to axiomatize and formalize probability theory. But he had missed the opportunity.

Two years before publication of "Grundbegriffe," Kolmogorov had written a piece titled "Über die analytischen Methoden in der Wahrscheinlichkeitsrechnung" (On analytical methods in probability theory). It was in this paper that Kolmogorov vindicated Bachelier. The aim of that paper was to apply differential equations—the analytical method to which the title referred—to so-called stochastically determined processes. These processes, nowadays known as Markov processes, predict the distribution of a system's state from knowledge of a previous state. Allow me to elaborate. Isaac Newton's mechanics is deterministic in that it allows the prediction of a body's future position and momentum given only its present position and momentum. Thus one is able to predict where a star will be one hundred years from now, just from knowledge about its state today. In contrast, stochastically determined processes allow the prediction of only a probability distribution. For example, given today's stock price, we may be able to predict for tomorrow no more than, say, "there is a 20 percent chance the price will rise, a 50 percent chance it will remain unchanged, and a 30 percent chance it will fall."

The states need not be discrete—up, unchanged, down; they could be continuous. After all, tomorrow's price could, for example, lie anywhere between $15 and $150.[11] Nor do the times for which predictions are made have to be discrete. Instead of today, tomorrow, the day after, one could make predictions hour by hour, minute by minute, or continuously.

And this is where Bachelier comes in. "Generally one examines only such models in probability theory in which changes in a system take place solely in a sequence of discrete moments in time," Kolmogorov remarks in the introduction to his paper. "As far as we know, Bachelier was the first person who systematically examined models in which probabilities vary continuously with time." Bachelier, at that time already sixty-one years old, could have expected nothing better, even if Kolmogorov did criticize his lack of rigor, as pointed out in Chapter 8.

The paper goes on to derive an equation that is known as the Fokker-Planck equation after the Dutch physicist Adriaan Fokker and his German colleague Max Planck of quantum mechanics fame. In independent papers in 1914 and in 1917, they had developed a partial differential equation that describes the time evolution of a variable's probability distribution. The variable could be a particle's position in a fluid, for example, or the distribution of a stock's price. Kolmogorov, unaware of the previous papers, rediscovered the method that had been developed by Fokker and Planck who, in turn, had rediscovered Bachelier's work, also without knowing it. (Regnault had anticipated the most important results of Bachelier's thesis of 1900, which preceded Einstein's, Smoluchowski's, and Langevin's findings by five years.)

Nowadays the Fokker-Planck equation is usually derived by making use of a tool that was developed, also independently, by the British mathematician Sydney Chapman and, once again, by Kolmogorov. The Chapman-Kolmogorov equation gives the probability of jumping from A to Z by investigating the probabilities for two consecutive jumps, say from A to P and then from P to Z, or from A to Q followed by a jump from Q to Z, and so on. It is like driving from New York in any direction and ending up in San Francisco. What are the chances of that happening? Well, you could drive via Dallas, Texas, via Wichita, Kansas, via Rapid City, South Dakota, or via any other route. The probability of ending up in San Francisco is then computed by adding the probabilities of all possible routes. This is what the Chapman-Kolmogorov equation does. The Fokker-Planck equation goes one step further. It develops the whistle stop tours into a distribution of probabilities of ending up anywhere on the West Coast, be it San Francisco, Santa Barbara, Los Angeles, or Tijuana, Mexico.

Actually, quite a few people studied the equation and everybody seemed to give it a different name. Kolmogorov referred to it as the "Bachelier case," while Lévy—emphasizing that he had never read anything by Bachelier—called it the "additive case." Meanwhile Bohuslav Hostinský (1884–1951), an underappreciated Czech mathematician, called it the equation of Smoluchowski, probably to show that he did not owe anything to Kolmogorov. Other mathematicians referred to it alternatively as the equation of Chapman, Kolmogorov, or Chapman-Kolmogorov-Smoluchowski. Chapman himself did not call it anything and was surprised when he first heard that his name had been attached to the equation.

But did such equations actually exist? Imagining an object and giving it a name—or several names, for that matter—is no guarantee that the object is real. For example, three-eyed Martian monsters can be imagined and even depicted in sci-fi comic books, but that does not mean they actually exist. So was the Chapman-Kolmogorov-Smoluchowski equation only a figment of Chapman's, Kolmogorov's, and Smoluchowski's imagination?

Kolmogorov did not think so and devised an ingenious strategy to ascertain that the equation that describes the probability that a particle jumps from one state to another is real. He first derived a so-called partial differential equation (usually abbreviated PDE), which has as its solution just the Chapman-Kolmogorov-Smoluchowski equation.[12] Then Kolmogorov turned the situation around: starting with the parabolic PDE, he pondered the question of whether it had a solution. If the answer was yes, then the solution could be none other than the Chapman-Kolmogorov-Smoluchowski equation. Hence it exists.[13]

By the late 1930s, it was generally known that under certain circumstances solutions to the parabolic PDE (a.k.a. the diffusion equation, a.k.a. Brownian motion) existed. The prerequisites that had to be fulfilled varied, with different authors requiring different conditions. Some were more restrictive, others less so, but the requirements were always stringent. For decades, attempts were under way to relax some of them. Unbeknownst to everybody, in the midst of World War II, the young German-French mathematician Wolfgang Döblin, whose tragic life story will be told in Chapter 12, made large strides ahead. But the reader will have to wait to hear that story because before that, our narrative moves to Japan.

ELEVEN

Accounting for Randomness

I HAVE ALREADY ALLUDED TO THE MATHEMATICAL TECHNIQUES of calculus, that is, of differentiation and integration as formalized by Isaac Newton and Gottfried Wilhelm Leibniz. Bachelier was the first mathematician to bring calculus to bear on problems involving probability, Kolmogorov the most recent. But there was a problem. Simple calculus deals with smooth mathematical functions that are, for that reason, aptly called differentiable. Brownian motion, however, the linchpin for the understanding of financial markets, is jerky and jiggly and nowhere near smooth. This is why it is said to be "nondifferentiable." In order to apply techniques from regular calculus to jerky processes, a new approach was required—stochastic calculus. The task to develop it fell on Kiyoshi Itō, our next pioneer in the search for the options pricing formula.

Born in 1915 in the Mie prefecture in southern Japan, Kiyoshi Itō studied mathematics at the Imperial University of Tokyo. As a student, he realized

that random phenomena are governed by statistical laws. But this was before Kolmogorov had put probability on a sound footing, and Itō was not
happy with the state of the field. "Although I knew that probability theory
was a means of describing such phenomena, I was not satisfied with contemporary papers or works on probability theory, since they did not clearly
define the random variable, the basic element of probability theory," he
wrote many years later. Itō, like his contemporaries, intuitively understood
that a random variable referred to measurement results of a random experiment. But the lack of a formal definition made it difficult to work with so
vague a notion. Like most mathematicians of his time, Itō regarded probability theory, as it was practiced then, as not quite serious. Upon graduation
in 1938, he joined the Cabinet Statistics Bureau. It must have been a cushy
job because—like Albert Einstein in the Swiss patent office—Itō had time
on his hands. Apparently the bureau's director realized that the young man
could do more important things than just compile statistics and he allowed
Itō enough leisure to study Kolmogorov's and Lévy's works in depth. During his five years at the Cabinet Statistics Bureau Itō wrote the paper that
would make random processes amenable to calculus. Entitled "On Stochastic Processes,"[1] the paper was published in the *Japanese Journal of Mathematics* in 1942, three years before Itō obtained his doctorate.

Before we consider Itō's contribution, let's return for a moment to
Brownian motion. Recall that microscopic particles suspended in a liquid
are constantly being pushed around by the liquid's molecules. Bombarded
from all sides in this manner, the particles' movements become so jagged
that the resulting paths are nowhere near smooth; not even the most minuscule stretch of the path is straight. This is what had bedeviled the Nobel
Prize–winning chemist The Svedberg when he tried to measure the particles' speed. All he wanted to do was to measure a portion of a particle's path
and divide this length by the time the particle needed to traverse it. But he
could never get a grip on the length. Whenever he zoomed in on the path
with a microscope, hoping to glimpse at least a small piece of smooth line,
new, more minute jags would become apparent. (See Chapter 6.)

The German mathematician Karl Weierstrass (1815–1897) had envisioned a theoretical example of such a path half a century earlier and devised the appropriate function. Defined as the sum of an infinite number of trigonometric expressions, it was henceforth known as the Weierstrass function. The sum does not explode toward infinity. To the contrary, by giving a definite, and finite, value at every point, the function is well defined. Furthermore, for points that are close, the values are close, so it is also continuous. But the function is not smooth. Even though it makes no sudden jumps, it is nowhere smooth. Wherever and however closely one looks, jags and more jags appear. With this, Weierstrass anticipated the paths of Brownian motion long before anybody ever thought of analyzing them mathematically.[2]

Now back to Itō. Ever since Isaac Newton discovered the laws of motion in the seventeenth century, the temporal behavior of physical objects—planets, cannonballs, bodies pushed around by other bodies—has been described by so-called differential equations. Based on current positions and speeds, these equations allow the computation of positions and speeds at any time in the future as, for example, when astronomers compute the exact orbits of planets a century or a millennium from now. The subjects of Itō's investigations were not physical objects but variables—the size of a population, the price of a stock—whose developments depend on not only the elapsed time but also random events. In principle, the behavior of such variables can also be described by differential equations. But there's a hitch.

The part related to time is easy. It is called the drift, or trend, and its effect can be predicted just like the future position of a heavenly body. One simply takes the equation's component that describes the variable's behavior over time, cuts the time period into narrow slices, computes the increase or decrease in each slice, and adds them up. For example, to see by how much a bank deposit, paying a fixed rate of interest, grows over time, we may divide the period into months, compute the growth in each month, and add everything up. We saw in the previous chapter that this is what Riemann and Lebesgue proposed.

In the short term, however, it is not the trend that dominates but random events (engineers call them collectively noise), and this is where the problem arises. Randomness in a variable is produced by the behavior of hidden underlying factors: the weather, supply and demand, cosmic rays, you name it. These factors constantly jitter and jag along Brownian paths. And while they jitter and jag, the variable itself has many opportunities to increase or decrease. In fact, whenever the paths make a kink, the variable's value may inch up or down, as in the case of a sudden hail storm, for example, that influences the price of wheat. And as The Svedberg had discovered to his horror, the closer one looks at Brownian paths, the more kinks appear. Thus, when cutting these paths into narrower and narrower slices, the number of kinks and, simultaneously, the opportunities for the variable's expansion or reduction, keep on growing. Hence the variable's future value could lie anywhere between minus infinity and plus infinity.

The inescapable conclusion is that there is no suitable way to slice the paths. Unless a different way of dealing with the situation can be found, we will have to forget about forecasting even a probability distribution.

Itō's way out of the predicament was the so-called Taylor expansion. Brook Taylor, an English mathematician in the early eighteenth century, discovered a method to approximate the numerical values of mathematical functions. With the help of Taylor expansions, we do not need to manipulate unwieldy mathematical expressions to analyze them but can use the much simpler approximations instead. Let me illustrate. The number 2 can be approximated by adding one, plus one-half, plus one-quarter, plus one-eighth, and so on. By adding more and more terms, you can get as close to the desired number 2 as you like. Adding the first ten terms, for example, gets you to within one-fifth of 1 percent of the number 2. On the other hand, if a ballpark estimate is good enough, you can make do with fewer terms. If you only want to get to within 10 percent of the number 2, you can stop after five terms.

The Taylor expansion (a.k.a. Taylor series) allows such approximations not for numbers but for functions. By computing and then adding a suitable number of terms, even the most intricate mathematical expression can be approximated. Usually the fit is so good that one or two terms suffice to get a qualitative grip on the function. All other terms can then be disregarded.

Itō regarded the function that described a variable's behavior as a combination of the effects of time and of random noise. So he expanded the function into a Taylor series and investigated the terms of the series. He quickly realized that only the first terms were of importance. All other terms could be neglected, in the same manner that one would neglect $\frac{1}{64}$, $\frac{1}{128}$ ... in the approximation to the number 2. In fact, only three terms were required to describe the variable's behavior. The other terms could be dropped without hurting the approximation. The resulting stochastic differential equation, with its three innocent-looking summands, is nowadays called Itō's lemma.

For regular differential equations that describe the movement of stars or the flight paths of cannonballs, two terms often suffice. The extra term in the stochastic differential equation is due to the jaggedness of the paths of the Brownian motion.

Once Itō had obtained a description of the variables' behavior, he was ready to forecast their future probability distributions. With the establishment of his lemma, this no longer presented a problem because the three terms of the Taylor expansion not only looked innocent but were innocent. Consequently the time period could be sliced without any ill effects, shifts within the slices could be computed and added up, and *voilà*, we had our forecast. Thanks to Itō's calculus, the behavior of variables that depend not only on time but also on random factors could now be quantified. Henceforth, their distributions could be predicted to a better degree than simply "between minus infinity and plus infinity."

There is one more thing. Variables that follow Brownian motion have the nasty habit of occasionally dropping below zero. One can visualize this by following the path of a particle suspended in liquid. Define any line as level zero, and the path is bound to cross it sooner or later.[3] But the prices of stocks and options, which is what this book is about, cannot fall below zero. You may lose your shirt on the stock exchange, but never more than your shirt . . . unless, of course, you take out loans in order to buy securities. So we are faced with a problem. Since the prices of securities cannot become negative, customary Brownian motion cannot quite serve as a model for them. An adaptation must be found to describe the behavior of stocks. "Geometric Brownian motion" consists of expressing price changes not in absolute dollar terms but as percentages.[4] Thus even if a stock, initially worth $100, falls twenty times in a row by 10 percent—an extremely rare event—it will still have a positive value ($12.16, to be exact). A further advantage of geometric Brownian motion is that the price changes in percent are what investors usually relate to in real life.[5]

Itō's path-breaking paper received little attention at first. Japan was isolated from the rest of the world during and immediately following World War II, and so this important work remained unknown to the mathematics community outside of Japan for almost a decade. But eventually word of his achievement filtered out. By 1954, when Itō was invited to give a series of lectures at the Institute for Advanced Study in Princeton—home to Albert Einstein, John von Neumann, Kurt Gödel—his methods were already well-known in the West.

Itō was named professor at Kyoto University in 1952 and remained there until his retirement in 1979. But he continued contributing to mathematics even as an emeritus. For his numerous profound achievements, he was elected to the Academies of Science of the United States and of France, was conferred honorary doctorates, and was awarded numerous distinctions like the Wolf Prize, the Kyoto Prize, and the Gauss Prize. But there is more. One of the greatest honors in mathematics is to become immortalized by having one's name associated with a mathematical concept. It is rare for an important lemma or theorem to be named after a mathematician. It is rarer still for it to be attached to a whole toolbox. But this is exactly what happened in Itō's case. The method of handling stochastic differential equations is nowadays known as the "Itō calculus."

Upon receiving the prestigious Kyoto Prize in 1998, a Nobel-level honor for work in basic science, advanced technology, art, or philosophy, Itō described his academic development.

Ever since I was a student, I have been attracted to the fact that statistical laws reside in seemingly random phenomena. Although I
knew that probability theory was a means of describing such phenomena, I was not satisfied with contemporary papers or works on
probability theory, since they did not clearly define the random variable, the basic element of probability theory. At that time, few mathematicians regarded probability theory as an authentic mathematical
field, in the same strict sense that they regarded differential and integral calculus. . . . When I was a student, there were few researchers
in probability; among the few were Kolmogorov of Russia, and Paul
Lévy of France. . . . I attempted to describe Lévy's ideas, using precise
logic that Kolmogorov might use. . . . I finally devised stochastic differential equations, after painstaking solitary endeavors. My first
paper was thus developed; today, it is common practice for mathematicians to use my method to describe Lévy's theory.

Eight years later, in August 2006, he was awarded the Gauss Prize for
"laying the foundations of the Theory of Stochastic Differential Equations
and Stochastic Analysis. Itō's work has emerged as one of the major mathematical innovations of the twentieth century and has found a wide range
of applications outside of mathematics. Itō calculus has become a key tool
in areas such as engineering, physics, and biology. It is at present of particular importance in economics and finance with option pricing as a prime
example." Of course, the "prime example" is the reason why Itō is so important for this book.

Kiyoshi Itō passed away in November 2008, at age ninety-three, in
Kyoto. His calculus set the stage for progress in the search for the options
pricing formula. But before we turn to these developments, the story of another man must be recounted, a tragic hero whose work would have remained unknown to the world for at least another generation had the
inquisitive interest of two French professors not brought it to light.

TWELVE

The Sealed Envelope

O

N MAY 18, 2000 (THE YEAR THAT THE INTERNATIONAL
Mathematical Union designated the World Year of Mathematics), a special committee of the Académie des Sciences
in Paris convened for a special task. Members were to decide on the fate
of a sealed envelope that had been deposited with the academy's secretary
sixty years earlier.

Once a year the committee assembles to decide whether to open the so-
called *plis cachetés,* sealed envelopes deposited by someone a century previously some for confidential safekeeping by the academy. Anybody,
scientist or nonscientist, Frenchman or foreigner, can bring or send a scientific paper and ask to have it kept as a *pli cacheté.* The pages of the paper
would be individually dated, signed by the secretary, and then kept hidden
in a box in the academy's archive. One of the earliest *plis cachetés* is from the
Swiss mathematician Johann Bernoulli in 1701, who was constantly in competition with his brother. The paper that he deposited dealt with a mathematical problem, and Johann stipulated that the *pli* would only be opened
after his brother, Jakob Bernoulli, had presented his solution to the same

problem.[1] Modern *plis* have contained, for example, advances in the theory of nuclear fission and the statistical theory of turbulence.[2]

The envelopes need not contain text but can also hold botanical specimens, designs for machines, or other paraphernalia. One early submission, dating from 1798, was a method of producing musical string out of twisted silk by an inventor from Versailles who sought to protect his design while awaiting a patent.

Not all *plis* remain concealed for a hundred years. Occasionally a depositor, or the next of kin, may request the opening of the envelope before the century is up. A submitter may also demand its withdrawal. But if there is no special request, the envelope lies unopened for a full one hundred years before anybody gleans its contents. Only then does the committee open it, designate one or more specialists in the scientific area to analyze it, and, if warranted from a scientific point of view, decide to publish a paper about it in the *Comptes Rendus de l'Académie des Sciences,* the prestigious proceedings of the academy.

Since the eighteenth century, scientists have used the sealed envelopes for a variety of reasons. A scientist may have discovered something significant without writing it up in a paper. If a competitor came up with the same idea while he himself was still working out the details, he could always demand the opening of the sealed envelope and prove that he had, in fact, been there first. The *plis cachetés* actually oppose the academic spirit in that scientific knowledge should be disseminated quickly in order to allow others to build on it.

In another sense, the *pli cacheté* is something of a cop-out. A scientist may have come up with a speculative theory. If it turns out to be correct, he may, at the opportune moment, pull the cat out of the bag. If he is wrong, then he may quietly demand the return of the envelope, unnoticed by malicious colleagues. Sometimes a scientist may feel that his findings are so revolutionary that the public is not ready for them. All too often, however, the system is used by pseudoscientists and quacks. Many *plis* have been opened that contained nothing but assorted hokum and obscure nonsense, and it may be safely assumed that the archives of the Académie des Sciences are full of sealed envelopes containing designs for perpetual motion machines, methods to trisect angles, and proofs that circles can be squared. Sometimes an unsuccessful but devious scholar may deposit an envelope just to garner interest, then demand its opening and request the publication of its content in an attempt to bypass the customary refereeing process. This has proven futile, however, since the committee strictly monitors the quality of the material that is ultimately published.[3]

In its session of February 26, 1940, the academy registered the *pli cacheté* 11.668. It had been sent by military post from the front lines in Lorraine in February 1940. World War II was raging; a German offensive was expected. The sender, a French infantry soldier of Jewish origin, was a radio operator in the French infantry. His name was Wolfgang Döblin.

He was born in Germany in 1915 but was granted French citizenship as a refugee in 1936. Döblin had recently received his Ph.D. in mathematics from the Sorbonne. His father, Alfred Döblin, was a noted Berlin neurologist and well-known author, most famous for his expressionist novel *Berlin, Alexanderplatz,* which describes the life of Berlin's small-time crooks and prostitutes in a poignant depiction of the times.

Alfred and his wife Erna had four sons—Peter, Wolfgang, Klaus, and Stephan. Though of Jewish origins, the family considered themselves to be Protestants, even before Jews came under persecution by the Nazis. Wolfgang, a voracious reader, demonstrated a precocious intellectual curiosity at an early age. By the time he was fourteen, he appeared on a Berlin radio show discussing big philosophical questions with his father, the famous novelist. In school he was known for his political interests. He made no bones about his socialist leanings—not an easy position to take for a youth in emerging Nazi Germany. The boy's relationship with his father was difficult; Alfred was a philanderer, and this made Wolfgang deeply protective of his mother.

When Nazism reared its ugly head in 1933, the Döblins realized that their professed Protestantism would not keep them from harm. Moreover, Alfred's fame put them at extra risk. On February 28, the day after the Reichstag fire, a police officer warned Alfred of his impending arrest.

Alfred immediately fled to Zürich by train. Erna and three of the sons followed four days later, while Wolfgang stayed on in Berlin for a few more months to finish his high school matriculation examination. With their possessions in storage, the whole family then moved to Paris. The residency permits they obtained did not allow them to work, however. Three years later, with the support of the diplomat André François-Poncet, the French ambassador to Berlin, the family was granted French citizenship. From then on, Wolfgang often used Vincent as a first name.

Wolfgang entered the Sorbonne to study mathematics and won two scholarships.[4] Obviously brilliant but standoffish to the point of arrogance, he was considered a loner by his colleagues. He spent his days studying and writing at the library of the Institut Henri Poincaré, his back always to the wall so that nobody could look over his shoulder.[5] Döblin took a liking to his fellow student Marie-Antoinette Baudot who, to his great chagrin, would marry the physicist Jacques Tonnelat and later become a physicist at the University of Paris. This was a great blow to him and in a film documentary sixty years later, his younger brother declared wistfully, "La femme qu'il aimait n'était pas pour lui." (The woman he loved was not meant for him.)

In spite of having attended none of France's *grandes écoles*, Döblin became an instant mathematical celebrity. His doctoral dissertation, written at the Sorbonne under the supervision of Maurice Fréchet, an eminent professor at the École Normale Supérieure, was awarded the *mention très honorable*, the coveted accolade that had eluded Bachelier. In the report on the thesis Fréchet remarks that "it is impossible to summarize Mr. Döblin's extremely numerous results here. . . . The richness of Mr. Döblin's results, the force which they reveal, permit this referee to judge his work as highly worthy of being presented as a thesis for the doctorate in mathematical sciences." By the time Döblin was twenty-three, he had published eight papers and another nine announcements in the *Comptes Rendus de l'Académie des Sciences.* He was a rising star in the theory of probability.

But then everything turned out differently. Döblin decided to join the French army. With a doctorate in mathematics, he could have easily obtained a commission as a reserve officer and avoided service at the front line. But as he told one of his brothers, by insisting on serving as a simple soldier he could show his gratitude to a country that had accepted him and his family at a time of desperation. One of his last scientific duties before his recruitment was a short talk, given at the session of the Academy of Sciences of October 24, 1938, about some astonishing mathematical results that he had recently obtained. Unfortunately, with the war looming, he never got around to actually writing the paper. (In the following year, a summary of his presentation was published in the academy's *Comptes Rendus*.) Barely two weeks later, on November 3, 1938, telephone operator Vincent Döblin marched out with his regiment.

It would be the first extended time away from his family for this sheltered young man. There is little record of his time in the army, but one can assume that as a Jew (even if baptized), an intellectual, and with an accent that constantly reminded his comrades of the enemy, not to mention his aloofness, he must have been somewhat ill at ease. But Döblin accepted his lot stoically. To account for his German accent he pretended to originate from Alsace, a part of France where German is used as a second language.

It was the time of the *drôle de guerre*. Also called "twilight war" by Winston Churchill, this phony war was a period in late 1939 and early 1940 when little happened on the western front. Following Germany's invasion of Poland, most Nazi forces were deployed in eastern Europe, while the French troops sat in their quarters, waiting for something to happen. The conditions in Givet, a small farming town in the department of Ardennes in northern France where Döblin's unit was stationed, were characterized as minor skirmishes. The soldiers spent their days exercising, practicing Morse code, and doing guard duty, their nights on haystacks in barns. Boredom was the biggest enemy.

Döblin rarely complained, however. Instead, he took up a fight, if not against the enemy, so at least against the gloominess of a foot soldier's day-to-day life. He turned to mathematics. "Fortunately, my mathematical work helps me fight despair. As I am not interested in alcohol, I do not have the luxury of getting drunk like the others," he would write in a letter to Maurice Fréchet, his professor in Paris.

Though he only had the rare hour to focus entirely on math, usually during night shifts hidden away in the telephone station that he was guarding and that provided some heat, his preoccupation with mathematical problems alleviated the dreariness and kept him from falling into a depression. This is when he began his work on Kolmogorov's equation, using a copybook that he had bought at the local greengrocer's. At first his arguments were not very explicit, the proofs not quite precise. But the more he worked, the more rigorous his work became.

———————

The battle of France began in earnest in May 1940. The German army advanced at an alarming speed, and on at least one occasion, Döblin was called on to repair telephone wires under fire, for which he was later awarded a medal.[6]

On June 14 the German army attacked the sector where Döblin's regiment was stationed, hammering the French lines with the toughest artillery fire the world had ever witnessed. By that evening, the French troops were forced to abandon their position and retreat on foot, carrying their arms and baggage during the night to avoid aerial attacks. The next day, the commander of the infantry regiment in which Döblin served was killed and replaced by a deputy. Completely exhausted, the regiment crossed the bridge over a channel and was immediately attacked again. The regiment's command post fell on June 19. Döblin's company, commanded by captain Renard, a Catholic priest, managed to escape immediate capture but was encircled by German troops. There was no hope of avoiding capture.

According to Renard's later testimony, Döblin was a model of courage and drive until that moment. But now, in the firm knowledge that the battle was lost, Döblin decided to try his luck on his own. On June 20, without telling anybody, he left his unit, possibly in an attempt to cross the German lines on his own. He marched throughout the night and arrived in the village of Housseras the next morning, amid hundreds of fleeing French soldiers. Exhausted from weeks of incessant combat, he realized that escape was all but impossible.

Döblin was determined never to fall into the hands of the Germans alive. If they found out, as they surely would, that he was the son of the decadent socialist writer Alfred Döblin, who was wanted by the Gestapo, not to mention that he was descended from Jews, the consequences would be unimaginable. Sometime earlier he had confided to the family of the farmer in whose barn he had slept that he was Jewish and that he would under no circumstances let himself fall into the hands of the Germans. He would rather shoot himself with the bullet that he always kept on him for this purpose.

Now the time had come. Without firing a shot, the Germans invaded the village of Housseras, to which Döblin had retreated, at about a quarter to nine in the morning of June 21. Döblin entered a farmhouse and burned all of his papers in the kitchen fireplace, not only his identity documents but also his scientific work; he could not bear the thought of the Germans getting hold of his scientific ideas. Then he went to the barn. Neighbors heard a shot ring out and when they came to investigate they found the body of an unknown soldier who had just killed himself. Döblin was twenty-five years and three months old. His remains were buried, unidentified, in the village cemetery. Four days later, on June 25, France capitulated.[7]

The family, who had managed to flee Paris in 1940 for America, thus avoiding the repercussions of the German invasion of France, heard nothing of their son's fate for the next five years. It was only in 1945 that a letter from Marie-Antoinette Tonnelat-Baudot reached them in their American exile. The woman who, years earlier, had spurned Döblin's advances, had spared no effort to find out about her former suitor's fate. She had been able to identify him on the basis of a metal bracelet that was found on his body. As soon as Erna saw the word suicide in the letter, she broke down. In April 1944 Döblin's exhumed remains were reinterred in the cemetery of Housseras.

After learning of Wolfgang's death, Erna collected her son's scientific papers and gave them to the archive in Marbach, Germany, where her husband's papers were collected, and to the archive of the Académie des Sciences. She was not aware of the *pli cacheté*, housed in the same building. In 1957 Wolfgang's father Alfred died. He was buried next to his son in the cemetery of Housseras. Erna took her own life only three months later and was buried, according to her last wish, next to her son and husband.[8]

Three months before his death, on March 12, 1940, Döblin mailed a letter from the front line to Fréchet. During the past years, he had sent nearly two dozen letters to his former teacher, ten from the front lines. Usually he told the professor a little bit about his life as a soldier before reporting about his work. Fréchet was always addressed with a respectful "Monsieur le Professeur" and the letters always ended with "Je vous prie, Monsieur le Professeur, de bien vouloir agréer mes sentiments très respectueux," which can be freely translated as "Sincerely yours" but has a much nobler ring in French.[9]

The letter of March 12 was one of the last Döblin would ever send. He had just finished writing up some ideas that he had carried with him before he joined the army but had not turned into a publishable paper. Now, at the front line, there was even less chance to do so. Fortunately, he had some free time during the preceding month and was able to at least sketch his ideas. But now, with the looming German attack, this period of relative quiet was over.

Döblin began the letter by describing his harsh living conditions. After an alert in January over ostensible German plans to invade Belgium, Döblin's unit was stationed in a village of about 150 inhabitants in the Lorraine province on France's northeastern border. The troops spent several weeks quartered in a freezing barn that let in snow. After reporting about life at the front line, Döblin came to the point. "About a month ago, I finished my article about the equation of Kolmogorov or, rather, at that point I had simply had enough of Kolmogorov's equation and just ended the paper. I sent a rough draft to my home, where it arrived, and the manuscript itself to the Académie des Sciences as a *pli cacheté*; but I am afraid they may not have received it." This was the only indication of the sealed envelope's existence.

For many years, nobody was aware of the *pli* that was gathering dust on a shelf in the academy. The only person who could have known about the sealed envelope was Maurice Fréchet. Unfortunately, Fréchet was at that time dealing with a personal tragedy: his wife had been run over by an

American Jeep. Consequently he forgot all about the *pli cacheté*. After the professor's death, his family handed over his papers to the archives of the Academy of Sciences. Fréchet had been a compulsive collector who never threw anything away. An abundance of material appeared: copybooks from the courses he took as a student, manuscripts of his publications, lecture notes, and all letters that had been addressed to him by dozens of colleagues throughout his long career. Among the latter were about twenty that Döblin had written. One of them was the note, written on March 12, 1940.

In 1991, a little more than fifty years after Döblin's death, a conference was to be held in his honor. Even though not a household name, he was nevertheless remembered among the cognoscenti for his notes in the *Comptes Rendus de l'Académie des Sciences* and the dozen or so important papers that he had published during his short lifetime. Preparing an address for the occasion, Bernard Bru, a historian of science at the René Descartes University in Paris, sifted through Döblin's correspondence with Maurice Fréchet. In doing so, he came across the letter from the front line in which Döblin tells his former teacher about the *pli* that he had sent to the Académie des Sciences in Paris for safekeeping. Bru was immediately intrigued. What did the envelope contain, scribblings of no particular value or a treasure trove?

For the time being, he had to curb his curiosity. According to the academy's rules the world would have to wait until 2040, a century after the deposit, for the *pli*'s content to be made public. The only way to open an envelope earlier was to get permission from the sender's next of kin. Bru contacted Wolfgang's octogenarian brother Claude, who had survived the war in Paris, in spite of having been hunted by the Gestapo and arrested twice by French collaborators. Claude agreed to the opening of the sealed envelope. At the next annual meeting of the committee charged with the supervision of sealed envelopes the request was considered and—there not being any reasons to the contrary—it was agreed to accede to it. On May 18, 2000, nearly a decade after the *pli*'s existence had been discovered by Bru, the secret would be lifted.

Bru and his colleague Marc Yor, a probabilist at the Pierre and Marie Curie University whom he had coopted for his endeavor, had been hoping for this moment for a long time. Now they waited with bated breath as the members of the committee ceremoniously unsealed the envelope in the chambers of the Academy of Science's archives. The envelope contained a simple notebook such as is used in primary and secondary schools throughout France. A blue photograph of a hill overlooking a village adorned its front cover, on the back the price of 1 franc and 50 centimes was printed. As could be expected, it was dense with mathematical equations. In the preamble, Döblin had noted, almost apologetically, "This manuscript was written in military quarters between November 1939 and February 1940. It is not quite complete and its presentation reflects the conditions under which it was written."

The text was written alternately in blue and in blue-black ink. Döblin used a second notebook, of the same type, as scratch paper in order to work out his proofs because some loose pages were inserted into the *pli*. Once satisfied with his demonstrations, he copied the text into the booklet with the photograph of the hill. There may have been a third notebook with a copy of the text. Döblin probably sent it to the Academy of Sciences as a backup. (Recall that he had told Fréchet that he feared his *pli* had not been received.) The academy's register of sealed envelopes shows the receipt on March 13, 1940, of another mailing by Döblin, but the envelope was never found.

In the notebook Bru and Yor found proofs of the results that Döblin had announced at the last session of the Academy of Sciences that he had attended. Unfortunately, as had to be feared with the author writing from the front line, the *pli cacheté* did not contain a completed, ready-to-print paper. Here and there words, phrases, or whole passages were struck out and rewritten. Some statements were erroneous and some were unreadable. Bru and Yor had to expend considerable effort to reconstruct the paper according to Döblin's intent.

The result was astounding, even for Bru and Yor, who had hoped for such a discovery. The subject treated in Döblin's booklet was the random movement of a particle in a medium. The Chapman-Kolmogorov equation, which describes the probability that a particle jumps from state x to state y during a certain time period, served as the starting point. As noted in Chapter 10, stringent prerequisites were needed in order to prove the existence of a solution to Kolmogorov's parabolic partial differential equation that describes Brownian motion. And this is where Döblin's genius became apparent. As Bru pointed out in an article celebrating the opening of the *pli,* Döblin managed to relax the prerequisites, thus proving that the Chapman-Kolmogorov equation exists under less stringent conditions than was hitherto believed. He then went on to derive properties of diffusion and of Brownian motion that were rediscovered some twenty years after his death, generally under more restrictive hypotheses. Several other results found in the copybook represent the state of the art even today.

The most important result contained in the *pli*—in Yor's opinion it alone would have justified the interest in it, even sixty years after its writing—can be found under the heading "transformation of variables" on pages 34 and 35 of the booklet. Here Döblin provides, in effect, a prototype of Itō's formula heralding the stochastic calculus.

As Kiyoshi Itō was ceremoniously inducted into the French Academy of Sciences in 1989, the *pli cacheté* still rested, its seal unbroken, in the dusty catacombs of the same institution. Only insiders who were familiar with Döblin's papers and announcements in the *Comptes Rendus de l'Académie des Sciences* were aware that the young man had ever existed. Nowadays Döblin is considered a trail blazing pioneer of probability theory, on an intellectual par with Nikolai Kolmogorov and Paul Lévy, prevented from fully developing his powers only by his untimely death. Decades later, Itō and others would tread the paths of stochastic theory, ignorant of the fact that these paths had been broken long before by the unknown French soldier who preferred to die by his own hand rather than fall into the hands of the enemy.

Döblin's accomplishments in his short life are truly remarkable. In barely five years, he published thirteen research papers, not counting the *pli cacheté,* and thirteen announcements in the *Comptes Rendus de l'Académie des Sciences,* most of them fundamental to the evolving theory of probability.

Lévy said this about his former protégé: "One can count on one hand the mathematicians—since Abel and Galois—who died so young and left such an important oeuvre."[10] Had Döblin's achievement become known in the early 1940s, or at least immediately after the war, the options pricing formula might have been developed decades sooner. Now that accountants, physicists, chemists, and mathematicians working in Europe, Russia, and Japan have laid the groundwork, the action turns toward America, where economists would take up the pursuit of the options pricing formula.

THIRTEEN

The Utility
of Logarithms

D URING THE NINETEENTH CENTURY AND THE FIRST HALF
of the twentieth, the discipline of economics looked at history for guidance, describing phenomena and events in narrative terms. Economic theory consisted mainly of anecdotes, rule of
thumb, and the manipulation of accounting data. Wary of using mathematics as a language, early economists painstakingly put their arguments
into words, formulating complicated explanations that could have been
described more simply through symbols and equations. Consequently
the discussions were vague and often futile.

Today economics uses mathematics as a language. Mathematics is precise and makes it clear when a point has been made and an argument
proven. When economists were forced to use mathematics to make problems and their solutions explicit, the eternal debates were temporarily
ended.[1]

The crucial turn took place in the decade after World War II, when the American economist Paul A. Samuelson determined that his field needed to be treated more rigorously. Writing in 1947, he complained that "the laborious literary working over of essentially simple mathematical concepts such as is characteristic of much of modern economic theory is not only unrewarding from the standpoint of advancing the science, but involves as well mental gymnastics of a peculiarly depraved type."[2] In his embrace of mathematics, Samuelson may be compared to Galileo Galilei, the seventeenth-century natural philosopher who used mathematics to explain natural phenomena. Before Galileo, mathematics was thought to deal only with abstract notions, having no relevance to the real world.

Samuelson infused rigor into a discipline that had at times become stale from never-ending arguments, with proponents and opponents alike talking in circles or arguing the same side without being aware of it. Starting with his doctoral dissertation, he showed that practically all economic behavior can be understood as maximizing or minimizing something, subject to constraints. For example, a family strives through its purchases to maximize its well-being, subject to the budget at its disposal; a company endeavors to maximize sales subject to the limits of its production facilities. Once a question in economics is precisely stated, it can be answered with existing mathematical tools, or new tools can be developed to make the problem solvable.

Two months older than Wolfgang Döblin, Paul Anthony Samuelson was born on May 15, 1915, to a Jewish family in Gary, Indiana. His father was a pharmacist. How different the American Jewish boy's life would be compared to the German Jewish youth's sad fate. Samuelson witnessed the Great Depression firsthand, but his life was happy, filled with professional success and many honors. When he was eight, his family moved to Chicago, where he went to school, graduating from the University of Chicago at age twenty. After an additional year to obtain a master's degree, Samuelson went to Harvard for his Ph.D.

At Harvard, Samuelson became interested in the thermodynamics of Josiah Willard Gibbs (1839–1903). One of America's greatest scientists of the nineteenth century, the famous Yale physicist had perfected the methods of classical thermodynamics and passed them on to his sole protégé, the polymath Edwin Bidwell Wilson. Wilson, whose interests spanned mathematics, physics, statistics, engineering, astronomy, and biology, was professor of vital statistics at Harvard's School of Public Health when Samuelson began his thesis. The young man now became Wilson's sole protégé, and his dissertation was greatly inspired by Gibbs's methods. Samuelson explicitly acknowledged the influence of classical thermodynamics on his thinking, and so its impact on the formation of economic theory is a matter of history.

Although by all accounts Samuelson had been a brilliant student, in 1940 he was denied a junior faculty position at Harvard due to blatant anti-Semitism. Harold Burbank, the chairman of the economics department, was well-known for his anti-Jewish views and blocked Samuelson's appointment. Samuelson's backer, Joseph Alois Schumpeter, a former Austrian finance minister who had taught in Germany and Austria and had come to Harvard with the rise of Hitler, was aghast. According to one biographer, Schumpeter proclaimed in a loud voice after the stormy faculty meeting during which Samuelson's appointment was discussed, "I could have understood if they didn't want to hire him because he is a Jew. But that wasn't it—he was just too brilliant for them!"[3]

It was Harvard's loss. Samuelson accepted an offer from the institution down the road, the Massachusetts Institute of Technology, and never looked back. Economics had been a neglected field at this engineering school, and over the years Samuelson made the department at MIT into one of the world's most important centers for economics teaching and research. It is still consistently ranked as one of the top economic departments.

Among the massive corpus of Samuelson's written work (vols. 1–5 of his *Collected Scientific Papers* alone comprise 400 articles written before 1976)[4], two books, published at the beginning of his career, stand out. The legendary undergraduate textbook *Economics: An Introductory Analysis,* first published in 1948, became the best selling economics textbook for the rest of the century, and beyond. By the time the nineteenth edition appeared in 2009, it had been translated into forty-one languages and had sold over 4 million copies.[5] Harvard University Press published Samuelson's Ph.D. dissertation as *Foundations of Economic Analysis* a year before the textbook was released. Chock-full of mathematical models, every page bristling with equations, it put the theory of economics on a rigorous mathematical basis. For many years it was a bible, albeit for the more advanced students of economics. What *Economics: An Introductory Analysis* was for undergraduates, *Foundations of Economic Analysis* was for graduate students.[6]

A reviewer hailed the book as the culmination of a great tradition. Soberly, Samuelson pointed out that such praise is a backhanded tribute in the dynamic field of science. It is far more important to be seminal and path breaking, he wrote, "to look forward boldly even if imperfectly," than to put the finishing touches to an established tradition. But he was being too modest. Samuelson advanced every subfield of the discipline and spawned several new ones, contributing fundamental insights to consumer theory, welfare economics, international trade, general equilibrium theory, dynamics, macroeconomics, and . . . the theory of finance.

Two highlights of Samuelson's career were winning the John Bates Clark Medal in 1947, awarded biannually by the American Economic Association to an "American economist under the age of forty who is adjudged to have made a significant contribution to economic thought and knowledge," and the Nobel Prize for economics in 1970, the second ever awarded.[7] There were many other honors in between and afterward. To quote the Nobel laudation by Assar Lindbeck from the Stockholm School of Economics: "More than any other contemporary economist, he has contributed to raising the general analytical and methodological level in economic science. He has in fact simply rewritten considerable parts of economic theory." Curiously, finance was nowhere mentioned in the citation. Lindbeck underlined how Samuelson recognized the advantages of strict formalization of economic analysis, thereby setting the style for the following generations of economists. He specifically mentioned four areas in which Samuelson's work had been seminal: dynamic theory and stability analysis, consumption theory, general equilibrium theory, and capital theory. Finance? Not a word!

In fact, finance theory came to Samuelson relatively late in life and he did his best work on the subject after he was fifty.[8] Given Samuelson's vast output of papers, it is maybe not surprising that the Nobel committee did not deem it necessary to take account of his then still young contributions to finance.

It was Samuelson who retrieved the work of Louis Bachelier from oblivion. He first heard of Bachelier in the mid-1950s when he received a postcard from Leonard "Jimmie" Savage, a statistician then at Yale, with an inquiry. Savage had stumbled on Bachelier's 1914 *Le jeu, la chance et le hasard* (Games of gambling, chance, and risk) in the library and was captivated. He sent postcards to several colleagues, asking whether anybody was familiar with the Frenchman and his work. One of the cards was addressed to Samuelson. Samuelson, who had started to become interested in the workings of financial markets, was interested enough to search for the book in the MIT library. He couldn't find it, but he found something even better: Bachelier's Ph.D. dissertation. Reading through the French text, he quickly realized its importance, at least for the historical record. "He seems to have had something of a one-track mind," he remarked and quickly added, "but what a track!" Bachelier's thesis, so far ahead of its time that it was largely ignored when first published, was finally starting to gain its well-earned, if belated, status as an outstanding contribution to economics.

In Samuelson's opinion, Bachelier's methods dominated Einstein's in every way. "The rather supercilious references to him, as an unrigorous pioneer . . . by more rigorous mathematicians like Kolmogorov, hardly do Bachelier justice. His methods can hold their own in rigor with the best scientific work of his time, and his fertility was outstanding." Bachelier had recognized before most of his academic contemporaries that prices on the stock exchange fluctuate randomly, rising and falling like coin tosses. Nevertheless, Samuelson remarked, speculators attempted to develop systems to outwit the market, like gamblers trying to outguess the game of roulette.

Samuelson distinguished between two kinds of investors. There were the "chartist-technicians" who try to infer patterns from the charts of past prices that would enable them to predict future prices. He held them in very low esteem "because they usually have holes in their shoes and no favorable records of reproducible worth." On the other hand, the much more reasonable "fundamentalists" and economists try to forecast what is going to happen to the price of wheat, for example, by basing their predictions on, say, the weather, the price of fertilizer, or dietary habits. But even the fundamentalists were surprised, Samuelson recounted, when they found out that there was hardly any difference between historical price changes and random number series. Bachelier had known that all along.

We now come to the heart of Samuelson's involvement with Bachelier. In spite of his great admiration for the French mathematician's work, he did not gloss over Bachelier's shortcomings. In fact, he dealt quite harshly with the inadequacies of Bachelier's derivations. The bone of contention was that Bachelier had made a crucial assumption that threw everything else into doubt. It was the innocuous-sounding postulation that price changes follow the so-called normal distribution, better known as the bell curve.

Since the time of Carl Friedrich Gauss, the nineteenth-century "prince of mathematics," the familiar bell-shaped normal distribution had become something of an icon. Practically everything that can be measured follows a normal distribution: men's height, women's weight, IQ test results, the errors in measurements. That this is so is guaranteed by the Central Limit Theorem, one of the cornerstones of statistics. The theorem says that when many independent random variables combine, the result is distributed according to the bell-shaped curve. A person's height, for example, is determined by many independent factors—father's height, mother's height, nutrition, environmental factors, and so on. When these factors are combined, they produce the heights that are normally distributed. Obviously many independent factors contribute to the determination of share prices. So nobody can really fault Bachelier for basing his mathematical treatment of price movements on the Gaussian distribution.

The mean of the normal distribution defines the middle of the bell-shaped curve. If observations of a feature, say the grades on SAT or GRE exams, are normally distributed, then half lie to the left of the mean, half to the right. Slightly more than two-thirds of all observations fall within one so-called standard deviation on both sides of the bell's center. (This is just about where the bell starts to curve outward.) Ninety-five percent of all observations fall within two standard deviations to the right and to the left of the mean; only 0.3 percent of the observations lie in the distribution's tails (i.e., farther away from the center than three standard deviations).

Since Bachelier assumes that price changes are drawn from a normal distribution, with the mean being zero, half of the price changes are negative. Recall an important fact about random walks and Brownian motion. As mentioned in Chapter 7, many drunken sailors who stagger around a lamppost eventually reach their homes, even if they are situated far away from the lamppost. Most significantly, this is true no matter whether the homes lie to the right or to the left of the lamppost. If the house lies to the left of the lamppost, a sailor's return to his home means that his leftward steps outweighed the rightward steps to such an extent that he was able to reach his residence.

Perhaps surprisingly, this implies that the price of a share on the stock exchange will eventually become negative. How so? Bachelier's model implicitly assumes that a gain or loss incurred one day is independent of the next day's gain or loss, or of the one after that, and so on. The only requirement is that gains and losses be normally distributed. Hence the share price could diminish by a dollar one day, by 50 cents the next day, by another dollar the day after, and so on. After a long streak of losses—a scenario that lies within the realm of normally distributed price changes—the accumulated losses could become greater than the share's initial value and the price of the share would become negative. In the same manner that some lucky sailor managed to stagger to his far-away home on the left side of the lamppost, negative price changes could, unluckily, accumulate until the share's price drops below zero.

What would it mean for a share to have a negative price? It would imply that the owner of the stock owes someone money. But can an investor get into debt at all? No, provided he invested his own money, he cannot. The key term is limited liability. In order to encourage investment, the stock exchange limits an investor's liability to whatever he invested. An investor can lose his entire investment, but the price of a stock can never drop into the red. Samuelson put it succinctly: "The GM stock that I buy for $100 today can at most drop in value to zero, at which point I tear up my certificate and never look back."

There is another way to look at the matter. As Bachelier, Einstein, and others pointed out, the distance traveled by an object that follows Brownian motion grows with the square root of time. Hence, by Bachelier's assumption, cumulative changes of share prices as well as option prices should also grow with the square root of time. On the upside this implies that options with very long times to maturity, say fifty years or a century, would command profits that could potentially tend toward infinity. On the downside it means that an investor could lose everything he invested initially, but never more than that. Hence potential gains are infinite but potential losses are limited. The conclusion is that options should command prices that become higher the longer the time to maturity, eventually tending toward infinity. But why would anybody be willing to pay an infinite amount of money for an option? Or why pay more than the value of the common stock itself?

Infinitely expensive options prices, negative share prices . . . what is going on? Well, what may be reasonable for drunken sailors and particles floating in a liquid need not be true for long-term commercial paper. Obviously Bachelier's assumption that price changes are normally distributed must be false. Samuelson concludes that "the absolute Brownian motion or absolute random-walk model must be abandoned as absurd." Note that he said absolute Brownian motion and absolute random walk. Why?

Exhibiting—in Samuelson's words—"a guilty awareness of the defect in his model," Bachelier apparently realized that he had a problem on his hands. With respect to the price of a share he wrote, "Nous supposerons qu'il puisse varier entre −∞ et +∞; la probabilité d'un écart plus grand que x_0 étant considéré a priori comme tout à fait negligeable." (We assume that [the price] can vary between -∞ and +∞; the probability of a drop greater than x_0 [the initial investment] may a priori be considered to be completely negligible.)

This justification is far too nonchalant. Waving aside the possibility of negative stock prices by considering them "tout à fait negligeable," in effect ignoring them, did not sit well with Samuelson. The subheading "Absurdity of Unlimited Liability" sets the tone of his critique. It demonstrates the illogical consequences that can arise if share prices that follow Brownian motion with normally distributed changes are assumed. Hence Bachelier's model needed to be adjusted.

What matters to an investor is not by how many dollars the price of a share rises or falls, but by what percent. Comparing a $1 increase in the price of Berkshire Hathaway's shares, which are worth over $100,000, with a $1 rise in the price of AT&T shares, which trade at about $20, would be misleading. It is not the absolute price change that matters, but the change relative to the purchase price.

Samuelson was not the first person to realize this. A high-energy physicist at the Naval Research Laboratory in Washington, D.C., by the name of M. F. M. Osborne had done so even earlier. In February 1958 Osborne delivered a paper at the lab's solid state seminar that was subsequently published in the journal *Operations Research*. It was titled "Brownian Motion in the Stock Market." In it he took a fresh look at the movement of share prices on the bourse.

Osborne's starting point was the Fechner-Weber law of psychophysics. Two nineteenth-century German scientists, Ernst-Heinrich Weber, a medical doctor, and Gustav Fechner, a psychologist, were the first to study the human response to physical stimulation. Investigating how humans react to increases in brightness, loudness, weight, and other stimuli, they found that the important factor was not the absolute increase but the relative increase in the stimulus. For example, a blindfolded man carrying a bag of rice to which more rice is gradually added will not notice any weight increase until the additional load passes a certain threshold. The interesting thing is that this threshold is proportional to the initial weight. If the bag of rice is heavy at the start, a higher increase is required for the carrier to become aware of the additional load than vice versa. Thus, if someone carrying a one kilogram bag of rice would notice the addition of, say, 100 grams, it would take the addition of a full kilogram for him to notice the extra weight in a ten kilogram bag. Drawn on paper with a logarithmic scale, the relationship between initial weight and the just noticeable additional weight is depicted by a straight line. Osborne thought the same should hold for wealth.

The hypothesis was quite old. Two centuries earlier, mathematician Daniel Bernoulli had already conjectured as much with respect to an individual's wealth. It had been, and still is, the only way to explain a paradox that was puzzling mathematicians at the time.

Daniel Bernoulli came from one of the most prominent scientific families the world has ever known. Three generations of Bernoullis had helped develop and advance differential calculus—with a little help from Isaac Newton, Gottfried Leibniz, and the Bernoulli family's compatriot Leonhard Euler—thus revolutionizing mathematics.

Daniel's cousin Nicolas Bernoulli had first raised the paradox in a letter dated September 9, 1713, to his French colleague Pierre-Raymond de Montmort. In their correspondence the two mathematicians considered the following game. A coin is thrown. If it comes up heads, the player wins a ducat; if it comes up tails, it is thrown again. If it now comes up heads, the player wins two ducats; otherwise it is thrown again. On the third throw heads is worth four ducats, and so on. Potential winnings double at every throw. But the probability that the coin falls tails up n times, and heads up only at throw $n+1$, is halved every time. The question that Nicolas Bernoulli asked is how much a player would be willing to pay in order to participate in this game.

The mathematical solution is surprising. Since the expected payout is computed as the probability of winning at each throw, multiplied by the amount to be won, the theoretically correct value of this game computes as: $\frac{1}{2} \times 1 + \frac{1}{4} \times 2 + \frac{1}{8} \times 4 + \ldots = \frac{1}{2} + \frac{1}{2} + \frac{1}{2} + \ldots$ This sums to infinity! But is this really what the game is worth to a rational person? Even the most ardent gambler would not be willing to risk all his money, and more, for the right to enter this game.

Correspondence about this problem between Nicolas Bernoulli, de Montmort, Daniel Bernoulli, and another mathematician, Gabriel Cramer, went on for nearly thirty years. In 1738 Daniel finally published his ideas on the subject in the paper "Specimen theoriae novae de mensura sortis" ("Exposition of a new theory on the measurement of risk") in *Commentarii Academiae Scientiarum Imperialis Petropolitanae* (Commentaries of the Imperial Academy of Sciences of Petersburg). Henceforth the predicament became known as the St. Petersburg paradox.

Bernoulli's crucial insight was that it is not money itself that must be considered when making a decision of this sort, but the utility that it confers on its owner. He then sought a mathematical function that describes the utility of money. He stated some requirements, or axioms, that a mathematical function must satisfy in order to qualify as a utility function. Obviously more money should give more utility. Hence the utility of money must always rise (axiom 1). But it must do so at a diminishing rate, since "a gain of one thousand ducats is more significant to a pauper than to a rich man" (axiom 2).

One mathematical function that satisfies these two axioms is the function "square root of x," which Cramer suggested as an example of a function that would well describe the utility money confers on humans. (This can be easily seen from the function's graph, which rises with increasing wealth but does so at a diminishing rate.) Bernoulli went a step further. He posited a third axiom, which states that "utility resulting from any small increase in wealth will be inversely proportionate to the quantity of goods previously possessed." This axiom—which is not universally accepted by economists today—would indicate that utility is described by the logarithm of wealth.[9]

Whatever the correct form of the utility function—be it the square root, the logarithm, or any other function that satisfies the first two axioms—

the additional utility that additional money confers on a person depends on his wealth. A millionaire should value a 1 percent increase in his wealth in the same way as a homeless person with $5 in his pocket would value a nickel. In a similar vein—and this was Osborne's starting point—it is not the absolute revenue in dollars that matters on the stock exchange but the relative gain or loss, expressed as a percentage of the initial share price.

According to Bernoulli, investors should judge the success or failure of their investments by first dividing the selling price by the initial price, and then computing the logarithm of the resulting number. Say a share was bought for $100 and you sell it for $100. Then the ratio of the selling price to the purchase price is 1.0, and the log of 1.0 is zero. You have broken even. Now you sell the share for $120. Divide the selling price by the initial price, which gives 1.2, and compute its logarithm, which is about 0.07918 . . . , a positive number. If the share price fell since the purchase was made to, say, $75, the ratio becomes 0.75 and the log of that number is –0.12493 . . . , a negative number. We see that positive log values denote gains, negative numbers denote losses, and a log value of zero indicates no change.

Osborne claimed that these log values, not the changes in the stock price, are normally distributed. He had a very good justification for his claim. Recall that normal distributions arise whenever many independent random variables are added. In a similar manner, the lognormal distribution arises when many independent random variables are multiplied with each other. And this is exactly what happens, since the monthly return on a stock can be calculated, for example, as the product of the daily returns. With daily price changes being random variables, and one day's return being independent of another day's, it is reasonable to expect changes in stock prices to behave according to the lognormal distribution. And, best of all, when Osborne checked the evidence from the New York Stock Exchange and the American Stock Exchange, the results bore out his claim.

Hence the changes in stock prices are lognormally distributed and it is this distribution, Osborne claimed, that guides investors in their decisions. It is also, he reminds the reader, "precisely the probability distribution of a particle in Brownian motion."

Osborne did not expect investors to explicitly compute logarithms of price changes and estimate the probabilities of their occurrence. "We do not claim that the trader sits down and consciously estimates the [changes in the logarithm of the prices] and [their probabilities] any more than one could claim that a baseball player consciously computes the trajectory of a baseball, and then runs to intercept it." However, he did believe that investors act intuitively according to logarithmic utility function and to lognormally distributed price changes. "The net result, or decision to act, is the same as if they did. In both cases the mind acts unconsciously as a storehouse of information and a computer of probabilities, and acts accordingly."[10]

While Osborne names the psychophysical laws of Weber and Fechner as the reasons for the results of his investigations, Samuelson's reasoning was firmly rooted in economic theory. In addition, Samuelson allowed stock prices to drift, that is, the mean increase in the stock prices would be some fixed, possibly small number, while Osborne assumed that the mean ratio of a stock's current price to its previous price should always be 1.0.

Whatever the slight differences between Osborne and Samuelson, the upshot of their findings was that price changes need to be expressed as a percentage instead of in dollars. This would resolve the absurd situation of negative stock prices that Bachelier's thesis had conjured up. With random, normally distributed price movements expressed in percentage terms, even a dozen successive downward jumps would not result in negative stock prices. The situation conforms to what can be observed on the stock exchange. A share whose initial value was $100 would still be worth $28 after twelve consecutive 10 percent drops, and $8 after two dozen drops. Had the price fallen by $10 each time, the share's value would have become negative after eleven drops. Thus price movements cannot be normally distributed in dollar terms, as Bachelier had thought. The data on stock exchanges around the world confirm that the changes in price expressed not in dollars, rubles, or euros but in percentages are, in fact, normally distributed.

Thus a share's gain or loss is best stated in a percentage, or as the ratio of the selling price to the purchase price. This can be well expressed by logarithms,[11] and the corresponding movement of prices on the stock market is said to follow *geometric* Brownian motion. Why geometric? Well, geometry deals with multiplication. For example, to compute the area of a quadrangle, its length must be multiplied by its width. So regular, arithmetic Brownian motion deals with additive changes in the share price, as when Thursday's dollar increase is added to Wednesday's dollar increase. Geometric Brownian motion, on the other hand, refers to multiplicative changes, as when the *ratio* of Thursday's price to Wednesday's price is multiplied with the *ratio* of Wednesday's price to Tuesday's price. Hence when price changes are expressed as ratios, or in percentage terms, we speak of geometric Brownian motion.

To summarize, why does Samuelson insist that geometric Brownian motion is a better model for price movements on the stock exchange than arithmetic Brownian motion? After all, the most important requirement, that tomorrow's value of a share depends only on today's, and not on past prices, is fulfilled by both models. But in contrast to arithmetic Brownian motion, share values that follow geometric Brownian motion are always positive, which agrees with the limited liability feature of company shares. Second, with geometric Brownian motion, price jumps, in percent, are equally distributed regardless of the stock prices. Third, the model is in accordance with human nature, as displayed by Weber and Fechner's psychophysical laws. Finally, and most importantly, the data observed on stock exchanges are consistent with price movements modeled by geometric Brownian motion. One problem that is unresolved even by geometric Brownian motion concerns the so-called fat tails, a problem that we will consider in the last chapter of this book.

FOURTEEN

The Nobelists

FROM THE MOMENT HARVARD DENIED PAUL SAMUELSON A position and the freshly minted Ph.D. chose MIT in its place, the latter institution became the place to be for all who wanted to learn, teach, and explore financial theory. MIT was home to three of the key figures of our narrative, Fischer Black, Myron Scholes, and Robert Merton, who spent much of their time in its department of economics or its school of business, or both.

The first of the three to beat the path to MIT was Fischer Black. Born in 1938, Black was an intelligent, curious child who started reading at age four and was already known in kindergarten as the smartest kid around. When he was ten years old, the family moved to Bronxville, just north of New York City, where no Jews or blacks were allowed to own property. The prejudiced, status-conscious townsfolk spent their workdays climbing corporate ladders and their free time playing golf and tennis and attending cocktail parties.[1]

As the scion of a WASP family, Black headed straight to Harvard for his undergraduate studies. He didn't even apply to any other college. Once there, he took a wide range of courses, including psychology, anthropology, sociology, mathematics, physics, logic, biology, and chemistry. Strangely missing in this eclectic selection was anything related to finance, although he mentioned to his parents that he was considering "even economics." He also experimented with psychedelics and—ever the scientist—took notes about his mental state every half hour. Once Black got arrested by the Cambridge police for demonstrating . . . but not against bigotry, racism, or world hunger. He was jailed for demonstrating against Harvard's decision to issue degrees in English rather than Latin. Behind bars, his bluster soon vanished and when a Harvard dean came to check on the detainee's conditions, Black quickly accepted the opportunity to get out.

As a twenty-year-old college senior, he married the daughter of a family even more WASPish than his. The girl was impressed by his nonconformist, intellectually challenging ways—not so her parents—and under his influence fancied herself something of a bohemian. A year later, a son was born. But the marriage did not last. The two young people separated and divorced a few months later. Black even gave up the right to see his son.

Toward the end of his third year in college, he finally settled on a major. He decided to do his B.A. in physics, took all the required courses, and graduated. In a letter to his parents he explained that he was not so interested in creating better things for better living, but rather in making better sense out of the physical world. (If the two goals could be combined, as in his later career on Wall Street, all the better.)

Black began his doctorate in theoretical physics at Harvard, but his interests soon wandered. Computers were all the rage, and he began thinking about using them to simulate higher cognitive functions. Then he took the course Automata and Artificial Intelligence taught at MIT by Marvin Minsky, the pioneer of artificial intelligence. He also took a summer job at the RAND Corporation where he met, and got to work with, Herbert Simon and Alan Newell from Carnegie-Mellon University.[2] Eventually he decided to change his field from theoretical physics in Harvard's Faculty of Arts and Sciences to applied mathematics in the Division of Engineering and Applied Sciences.

In spite of the change of direction, his doctoral work was going nowhere. In June 1962, after futile warnings by his professors that there would be negative consequences if his academic performance did not pick up, he was kicked out of Harvard. Entering the real world for the first time, he started working with Bolt, Beranek, and Newman (BBN), an engineering firm that provided consulting services in acoustics. Due to the heavy computations required for such work, the firm's business soon spread into computing and Black—who was still interested in artificial intelligence—got the task of writing a computer program that would answer questions using mathematical logic. When Minsky heard about his work, he offered to direct Black's doctoral work. Thus the disgraced student eventually returned to Harvard in a roundabout way to complete his thesis, informally under Minsky's supervision. It was titled "A Deductive Question Answering System." He obtained his Ph.D. in June 1964 and ended his involvement with artificial intelligence.

Now it was time for Black to look for a job that could turn into a career. He interviewed and got a position with Arthur D. Little (ADL) in North Cambridge, a consulting firm that did projects similar to the work Black had done at BBN, except that ADL's clients were usually businesses while BBN did most of its work for the government. At ADL, Black got his first taste of his vocation: finance. The person who whetted his appetite was his colleague Jack Treynor.

Treynor, a mathematics major from Haverford College with an MBA from Harvard Business School, had wondered since his student days about the relationship between an investment's riskiness and the appropriate discount factor. Obviously the discount factor should be higher the riskier the investment, but how high exactly? And how is risk to be measured? Treynor, who had joined the operations research department at ADL, spent many weekends thinking about the subject. His idea was to use the discount rate that prevails in the market to value investments. In August 1961 he wrote a paper summing up his thinking over the previous years. The paper made the rounds of some of the greatest minds working in economics: a colleague sent it to Merton Miller at the University of Chicago (Nobel Prize in economics 1990), who sent it on to Franco Modigliani at MIT

(Nobel Prize 1985), who suggested that Treynor meet Bill Sharpe, then at the University of Washington (Nobel Prize 1990). It raised the interest of all who read it. Modigliani, who had published a very influential paper on the subject together with Miller three years earlier, invited Treynor to spend the academic year 1962–1963 at MIT.

The crucial question now concerned the discount rate that prevails in the market. Building on Harry Markowitz's work on diversification and portfolio theory (Nobel Prize 1990), Treynor concluded that an investment's riskiness must be seen in relationship to the riskiness of the market as a whole. After all, most investors do not own shares from just a single company but usually hold a whole portfolio of risky assets. They do this because they believe, correctly, that investors should diversify their stock holdings among different companies, industries, and countries. Thus when the market for, say, real estate goes up, shares in telecommunications may fall, and while stock in France rises, South American shares may fall. By diversifying, investors keep their overall risk at an acceptable level. What Treynor came up with during his year at MIT was the capital asset pricing model (CAPM, pronounced "cap-emm," for short).[3]

According to CAPM, an asset's risk is composed of a diversifiable component and a nondiversifiable component. The former is the risk that is specific to the stock under consideration. The price of a share of AT&T, for example, may depend on technological advancements, the company's marketing strategy, the health of its president, and many other factors. This part of AT&T's riskiness can be "diversified away" by combining its shares with many others into a diversified portfolio. When part of a portfolio contains many different shares, the specific riskiness of the AT&T share plays only a minor role. Thus the diversifiable risk is that part of an asset's risk that is specific to it, over and above the risk that is inherent in the market as a whole. It is "diversifiable" in the sense that it can be alleviated by diversifying one's portfolio.

The nondiversifiable risk is called market risk, or "systematic risk," because it depends on the market's vagaries. Due to market risk, all stocks, including the one under consideration, tend to move upward in good times (bull markets), and downward in bad times (bear markets). This part of the risk is here to stay. No matter how much the investor diversifies his portfolio, it cannot be diversified away. But how much will it move up in a bull market or down in a bear market?

CAPM shows how a share—an asset in Treynor's words—is linked to the market. A parameter called "beta" is estimated, which measures the asset's sensitivity to, or correlation with, the market. The performance of the market itself is often measured by a stock index, for example, the one computed by Standard and Poor's or the Dow Jones Index.[4] A beta of 1.0 indicates that the asset varies with the market, and on average returns whatever the market returns. If the Standard and Poor's index rises by, say, 4 percent, the asset's return is expected to also rise by 4 percent. Assets with higher or lower betas would be expected to yield correspondingly higher or lower returns. A beta of zero indicates no correlation with the market at all. Hence the asset would, on average, return no more than the risk-free rate—the interest rate that one receives by depositing money into a savings account or by investing it in zero-risk Treasury bills.*

Specifically, CAPM says the following. If r is the risk-free rate of interest (e.g., the interest paid on savings accounts), and m is the market return (e.g., the growth of the S&P index), then $m-r$ needs to be multiplied by the asset's beta, and the result added to r. This is what the asset should return. For example, if the risk-free rate of interest is 2 percent, the yearly return on the Standard and Poor's index is 4 percent, and the company's beta is 1.5. Then the asset should return at least 5 percent. Looked at differently, given the return on an asset with a certain beta, the model determines the correct price of its stock. If a company's beta is 2.5 and its share price grows by $4, the share price should be $57. (The return should be 7 percent by CAPM, and $4 is 7 percent of $57.) CAPM also provided a clever tool for the evaluation of investment managers. Henceforth, managers would not be assessed simply by whether the returns they generated were high or low, but whether these returns were high or low, given their portfolio's beta.

Treynor concluded that the appropriate discount factor to be used for new projects and investment decisions by a company is the return on assets

---

* See footnote on page xiv.

as required by CAPM. In the above example, new projects of a company with a beta of 1.5 should yield at least 5 percent; with a beta of 2.5 they should yield at least 7 percent.

Back at ADL, Treynor befriended and started working with Fischer Black. It was the first time the new recruit, whose wide-ranging interests had hitherto covered physics, applied mathematics, psychology, and artificial intelligence, was exposed to the field of finance. Together they started working on financial problems for ADL's clients. One such case was Yale University. Its president fostered ambitious plans for the university's development, which needed to be financed from the proceeds of its endowment. The idea was to generate additional funds from the university's investments in the stock and bond market by taking on greater risk. ADL's assignment was to provide tools for the evaluation of the endowment managers. Yale was Treynor's last case at ADL. The consulting company did not appreciate the importance of his novel approach sufficiently, and Treynor left the company to join the investment banking firm Merrill Lynch. Fischer Black inherited his case load.[5]

At about that time, Black first heard about the so-called efficient market hypothesis. This hypothesis, propagated by the University of Chicago school of thought, states that all publicly available information about a certain company is immediately reflected in the price of its shares. As soon as news breaks, be it good or bad, prices adjust accordingly. By the time you read about it in the *Wall Street Journal,* the price has already taken the information into account. Hence in principle, there is no way to beat the market— to get more than the market return—unless you happen to have insider information. You could, for example, know relevant details before anybody else because you work for the company or because you received a tip from someone who does. But even if you know for sure that the research department has just found a cure for cancer or that a disgruntled customer is preparing a multimillion dollar lawsuit, you are not allowed to act on such knowledge. The trading of shares by "those in the know" is strictly forbidden. Laws have been enacted in order to level the playing field for all investors and not give undue advantage to insiders. After all, if people believed that only insiders were able to make money, nobody would invest. Hence investors convicted of insider trading routinely receive harsh prison sentences.

CAPM, the efficient market hypothesis, and normally distributed percentage changes in stock prices, as present in geometric Brownian motion, are all the ingredients that are needed to create nice models of financial markets . . . well, nearly. Similar to many other propositions in economics, the efficient market hypothesis is only true, if at all, under restrictive circumstances. There is a joke about a finance professor who ignores a hundred-dollar bill lying on the sidewalk. After all, he reasons, according to the efficient market hypothesis, if it had really been a hundred-dollar bill someone would have already picked it up. In reality, an investor may occasionally be able to identify market inefficiencies before other investors do. If markets happen to be out of whack for a short while, and they often are, alert traders are able to turn a profit if they are quick. But don't get your hopes up: usually such investment deals involve simultaneous long and short trades with a multitude of derivatives, carrying different expiration dates, in various currencies, intricately linked together in complex strategies.

Black was hooked. Financial theory was the intellectual challenge for which his quick mind seemed predestined. The subject would occupy him for the rest of his life. Unfortunately—for ADL, that is—his superiors at the consulting company failed to recognize Black's potential and refused to give him a raise. This was his chance to become independent and he promptly founded a consulting company of his own, Associates in Finance. The plural in the company's name was only barely correct because apart from Black and a part-time secretary, the firm had only one other associate, an up-and-coming finance professor from MIT, Myron Scholes, whom Black had met while they both consulted for one of ADL's clients.

Scholes, Black's junior by three years, was born in Canada in 1941. His father was a dentist, his mother a businesswoman who, together with her uncle, operated a chain of small department stores. She died of cancer a few days after his sixteenth birthday but had managed to instill in her son a penchant for business and economics. Scholes was a good student, always near the top of his class, even though his learning was impaired by a serious problem with his eyes. Finding it difficult to read, he learned to be a good listener and think conceptually. (At age twenty-six, he received a corneal transplant that improved his vision substantially.)

Scholes initially intended to study physics or engineering at McMaster University, but decided on economics instead. After graduating in 1962, he considered law school but then decided to follow his late mother's wishes and join his uncle in publishing. Before embarking on a business career, however, he insisted on attending graduate school at the University of Chicago. What was meant to be no more than an interlude before getting on with real life became the experience that would define his career.

After the end of his first year in graduate school, Scholes obtained a junior computer programming position at the university. There was a problem, however: the young graduate student had no notion of programming. This was the early 1960s and most people considered computers mysterious machines for the Pentagon or NASA. During his first days on the job, when professors came to the computer lab to ask for assistance, they found a junior programmer who fended them off by asking them to return when one of the senior programmers was available. Since no senior programmers ever showed up—there weren't any—he was forced to learn everything by himself from scratch.

Scholes quickly studied the rudiments of programming and fell in love with computers. In later years he described himself as one of the first computer nerds, working all hours of the day and night, furiously hacking away at teletypewriters, sorting punch cards, and waiting for printouts from time-shared computers. During this summer job, Scholes turned into a computer wizard. Had there been a computer science department at the University of Chicago back then, he might have become a computer scientist.

However, his first love was still economics. Future Nobel Prize winners Milton Friedman, George Stigler, and Merton Miller were on the faculty at Chicago, and brilliant students abounded. Scholes met and become lifelong friends with fellow students who would be among the most important names in economics and financial research in the coming decades. Meanwhile he continued to provide programming services to the faculty, and many of the famous professors were his clients. It may have been partly because Merton Miller did not want to lose the gifted programmer that he suggested that Scholes enter the Ph.D. program.

Scholes entered the young field of financial economics with full force. The topics of his interest were asset prices, the demand for traded securities (taking into account their risks and returns), the efficient market hypothesis, and how arbitrage prevents investors from earning abnormal profits. The topic of his thesis was a test of the efficient market hypothesis. He investigated the effect of information on share prices—for example, when an investor who may or may not be informed about something that has not yet hit the newswires suddenly sells a large parcel of shares. Upon completing his doctorate in the fall of 1968, Scholes was named assistant professor of finance at MIT's Sloan School of Management. He met Fischer Black at one of Franco Modigliani's Thursday evening finance seminars. Together they revolutionized the way securities are traded.

With a degree from one of the foremost economics faculties and a thesis that dealt with the functioning of financial markets, Scholes was well on his way as an up-and-coming scholar. He was offered a job with Wells Fargo Bank in San Francisco, which at the time was doing trailblazing work in finance. Its newly founded management sciences department was one of the first institutions to put the new methods of fund management to practical use. John McQuown, a mechanical engineer by training, had been hired by the bank to head the department and get its investment operations up to speed. He was to do so independently of the existing money management operations. Recognizing the potential of the new theories, McQuown was eager to implement them and build a practical business around it. The first step was to hire every specialist in sight, thus putting a number of Scholes's colleagues and former professors on the bank's payroll. Preferring an academic career, Scholes declined a job offer but joined Wells Fargo as a consultant.

Since Scholes did not have sufficient time to do all that the bank expected of him, he suggested also hiring Fischer Black. Wells Fargo followed the advice and employed Black's firm, Associates in Finance. The two complemented each other well, Fischer being the ideal person to tackle theoretical questions while Scholes dealt with the data. Scholes would often leave MIT at three o'clock in the afternoon in order to spend the remainder of the day at Black's office. Wells Fargo was the firm's most important client, accounting for half of the firm's billings. Black and Scholes had the time of their lives. It was just the combination of theory and practice—of academic research and real-world applications—that appealed to them.

But there were setbacks. Wells Fargo's brief for Associates in Finance was to design financial products that would make money. Basically the work consisted of testing CAPM and market efficiency. If CAPM did not always hold, how could market inefficiency be exploited to make profits? And above all, are there pockets of inefficiency hidden away somewhere, in spite of the efficient market hypothesis?

Black and Scholes spotted one. Examining historical data, they realized that stocks with low beta values, which indicate low correlation with the market, had higher returns on average than they should have had according to CAPM. The opposite held for stocks with high beta values. In other words, low beta stocks were undervalued, and high-beta stocks were overvalued. Wells Fargo's two consultants surmised that this inefficiency must have arisen because of borrowing constraints. Ideally, investors would have liked to leverage their holdings—borrow money that would be invested in order to bring their portfolios in line with their preferred risk level. They were prevented from doing so, however, because of the banks' traditional reluctance to lend money at an acceptable interest rate to individual customers. Black and Scholes were quick to design a portfolio around that opportunity. The Stagecoach Fund combined low-beta stocks and was leveraged by borrowing at a rate that no individual investor could obtain. It was designed to beat the market by a long shot.

Alas, it was not to be. At the crucial meeting, a senior associate from Wells Fargo's financial analysis department pointed out that low-beta stocks are closely correlated among themselves, which would make the fund vulnerable to downturns. To the dismay of Black and Scholes, the Stagecoach Fund was voted down. Furious, Black left the meeting.

A year before Scholes was hired as an assistant professor by MIT, a promising graduate student had joined its economics department. He had studied applied mathematics at Columbia University as an undergraduate and had begun graduate studies at the California Institute of Technology when he changed direction and decided to devote himself to economics. His name was Robert Cox Merton.

Merton's father was a famous figure in academia. Born to Jewish immigrants from eastern Europe in 1910, Robert King Merton grew up as Meyer Schkolnick in Philadelphia. He became one of America's foremost sociologists and historians of science. Phrases like "self-fulfilling prophecy," "focus group," and "role model" that have entered everyday language are attributed to him. He first taught at Harvard and, starting in 1941, at Columbia, which became his academic home for nearly four decades.[6] His second wife, Robert's mother, hailed from a southern New Jersey Methodist/Quaker family.

Robert was born in 1944. Together with his two sisters and two dozen cats, he grew up in Hastings-on-Hudson, a middle-class town of about 8,000 inhabitants outside New York City. Despite its small size and mainly blue-collar population, Hastings counted among its residents no fewer than five current and future Nobel Prize winners in physics, medicine, and economics, as well as sculptor Jacques Lipchitz. With a house full of books and by example, Merton Sr. became Merton Jr.'s role model for learning and scholarship. (Remember, he had coined the phrase.) The Hastings public schools provided high-quality education and Merton was a good student, even though he was not at the top of his class.

But school was not uppermost in the youth's mind. His interests included baseball and cars. Counting the days until he would be old enough to obtain a driver's license, he passed the waiting period handing tools to older auto buffs at stock car races. Among his other interests were money and finance. As a boy of eight or nine, he amused himself by creating in his mind fictitious banks. After his father introduced him to the stock market, Merton bought his first shares—General Motors—at age eleven.

Once he reached the legal age for driving, he spent his free time building street hot rods that he raced at drag strips in upstate New York and Long Island. Intending to become an automobile engineer, he entered Columbia University's engineering school in 1962. He took courses in pure and applied mathematics, engineering, sociology, and even registered for English literature and humanities, receiving no more than C or D grades in the latter. Nevertheless, his first academic article was a term paper for an English literature class on *Gulliver's Travels* that was published in the *Journal of the History of Ideas*. He also took an introductory course in economics, which used—what else?—Samuelson's *Economics* as a textbook, as well as evening classes in accounting and investments.

His interest in the stock market never waned. While at college, he hung around brokerage houses to watch the tape and trade. It was at about that time that he had his first experience, and a profitable one at that, with what would later become known as risk arbitrage. There were rumors among brokers that I. M. Singer, the well-known sewing machine maker, and Friden, a company that built calculating machines, were about to merge. Since the price of the shares of the new, merged company would be a weighted average of the old shares, the deal involved buying Friden shares, which were expected to rise, while selling short Singer shares, which were expected to fall, at a ratio of 1 to 1.75. Profits would be a sure thing, provided the merger took place. Luckily, a year later the two corporations did merge and Merton made a small bundle of money.

In 1966 Merton graduated with a degree in engineering mathematics, and one week later he married his high school sweetheart, a model and TV actress. Soon the couple was on the way to Pasadena, to the California Institute of Technology, where Merton was to pursue a Ph.D. in applied mathematics.

He soon realized, however, that his passion was not for mathematics. He liked the learning by doing approach practiced at CalTech, but his academic interests lay elsewhere. It was economics that had caught his fancy. Recent advances in macroeconomics made him, and many others, believe that it would soon be possible to control business cycles, limit unemployment, and stop inflation. The potential to affect the lives of millions appealed to Merton. Add to this his predilection for the stock market, and his choice was made. While at CalTech, Merton took to arriving at a local brokerage house at 6:30 AM, the opening time of the New York Stock Exchange, and spending a few hours trading stocks, options, warrants, and convertible bonds before classes started. At the end of the school year 1966–1967, he was awarded a master's degree in applied mathematics by CalTech and went back East to begin a graduate program in economics.

Merton had applied to several universities but only got into one: MIT. The decision would turn out to benefit both sides. Merton's first-year adviser recommended a course by Paul Samuelson in mathematical economics. Merton took the course and wrote a term paper good enough to be published. This so impressed Samuelson that he hired the twenty-four-year-old student as an assistant.

A problem that Samuelson had been thinking about for close to a decade was how to price warrants. Warrants are similar to options, except for some technical differences. A call warrant entitles the holder to buy a share of the company that issued it at a specified price and a specified date in the future. If the market price then is higher than the price stated on the warrant, the holder can make an immediate profit. If it is lower, he simply walks away. Put warrants correspond to put options in that they entitle the holder to sell a share at a specified price. Often warrants are added to a corporation's bonds as a sweetener to induce investors to purchase them.

The big question was what the price of a warrant or an option should be. How much should one pay for the right to buy or sell a share at a specified price sometime in the future? Obviously the price at which a warrant is traded in financial markets is determined by supply and demand. But Samuelson felt that there must be some intrinsic value around which the price should fluctuate.

An answer was suggested in the popular 1967 book *Beat the Market: A Scientific Stock Market System,* by Sheen T. Kassouf and Edward O. Thorp.[7] It sold like hotcakes. After all, who doesn't want to beat the market? The secret that the authors revealed in their book was to create hedged portfolios by buying stock while selling warrants short. The idea was sound, but overvalued warrants had to be found that could then be sold short. The prize question was how to spot warrants that were overvalued.

Amazingly, Kassouf and Thorp managed to obtain a formula that supposedly gave the warrant's "correct" value. They did not derive it from economic principles but—horror of horrors—by simply fitting a curve to actual prices. Given their improvised method of investigation, they could not prove the formula's correctness, of course. Except for the fact that it gave a good fit, they had no way of knowing whether their formula was correct. Unsurprisingly, Samuelson was not impressed. "Just as astronomers loathe astrology, scientists rightly resent vulgarization of their craft and false claims made on its behalf," he wrote in a review and concluded that "the book will make more money for its authors than any other use of its system could." Yet a few years later, Kassouf and Thorp's formula turned out to be remarkably close to the real thing. Like blind hens, they had hit upon a nugget. And they knew it. But they could not prove it.

Samuelson's own initial attempt to derive an equation that gives the warrant's price was not altogether convincing either. It did not take into consideration the investor's utility, an indispensable ingredient of any pricing formula, as the thinking went at the time. (See Chapter 13.) As Merton continued to seek a better pricing formula, his mathematical training, combined with his experience in trading stocks, bonds, options, and warrants, came in handy. Samuelson suggested that they try to work out the problem together. The result of their collaboration was a paper published in 1969, "A Complete Model of Warrant Pricing That Maximizes Utility." In it they specifically consider the risk preferences of an investor—a consequence of his utility for money—and derive the price of warrants based on the practical assumption that returns on stocks and warrants are determined by supply and demand.

Samuelson was to present the paper in October 1968 at the inaugural session of the MIT-Harvard Mathematical Economics Seminar, which was to become a staple of the economics profession in the Boston area. But as Merton put it many years later, "my co-author decided that I, the second-year grad student, and not he, the Institute Professor, would give our paper." It was a memorable occasion with the crème de la crème of the economics departments of Harvard, MIT, and other institutions in attendance. According to an account by Columbia University professor Perry Mehrling, Samuelson opened it with the following words: "This is a joint paper, and my co-author will present it. I'd like to introduce him as professor, but he is not a professor. I'd like to introduce him as Doctor, but he has no Ph.D. So I'll just introduce him as Mr. Robert Merton."

Meanwhile, Black, who was still consulting for Wells Fargo Bank and other clients, together with his associate Myron Scholes, continued hanging around MIT. Besides attending the occasional student seminar, he was a regular guest at Franco Modigliani's weekly seminar on finance. His contacts with Samuelson were less close. The professor had little appreciation for the banker-like man whose well-groomed, well-dressed appearance clashed with the academic environment at MIT.

Black had missed the inaugural session of the MIT-Harvard Mathematical Economics Seminar and met Merton for the first time a month later at a student seminar where Merton presented the gist of his thesis. After the talk, they chatted a bit but then lost touch again.

With the completion of his thesis in early 1970, Merton needed to get a full-time job. Samuelson nominated the promising scholar for the position of junior fellow at Harvard, but just like his own regrettable experience with Harvard three decades earlier, nothing came of it. Subsequently Merton applied for a teaching position at MIT's Sloan School of Business. By then, Scholes—three years older—was an assistant professor there, and one of the faculty members who interviewed the candidate. Merton was offered the job.

FIFTEEN

The Three Musketeers

I N THE 1960S, WHEN ROBERT MERTON WAS STILL A STUDENT and Fischer Black and Myron Scholes were just beginning their investigations, several other scholars attempted to find the price of an option. The first real progress since Louis Bachelier's Ph.D. dissertation and Vincenz Bronzin's booklet at the beginning of the century came when a graduate student in economics at Yale, Case Sprenkle, analyzed empirical data on warrants.[1]

Sprenkle modified several of Bachelier's assumptions, some of which advanced the state of the theory and some of which did not. First he assumed that stock prices follow a geometric, not arithmetic Brownian motion. As noted earlier, this assumption was more reasonable because it avoided the unintended consequence that stock prices could become negative. Moreover, it meant that price movements are not equally likely to jump up or down by $1, regardless of whether the stock's current price was $10 or $1,000.

Furthermore, unlike Bachelier, Sprenkle did not postulate that price movements offset each other on average. Rather, he allowed for a positive drift—a positive return on shares in the long run and thus positive interest rates. Next, he posited that investors possess utility functions and are risk averse. In fact, he was mainly interested in investors' risk attitudes, and he investigated warrants precisely because he thought their prices would reveal something about investors' risk attitudes. Unfortunately, in this he was wrong. As we will see below, the price of an option does not depend on the investors' risk attitudes.

He was also wrong in assuming that risk-neutral individuals would be willing to pay for an option regardless of its expected value. This would have sounded reasonable, except for one detail: Sprenkle brushed aside the fact that the option must be paid for on the day of purchase whereas any possible payoffs can be expected only upon the option's expiration. Thus, in this instance, he ignored the time value of money, even though his model did include interest rates.

A year later, in 1962, James Boness, a graduate student at the University of Chicago, went a step further than Sprenkle. By discounting the stock's terminal value to the present, he allowed for the time value of money. His opening remarks reflect the desolate state of his chosen area of research. "Securities markets, for all the publicity given to them and for all the interest they seem to command, have not until recently received the attention of economists . . . Nor has the decision-making procedure of investors in equities been as well documented and analyzed as those of consumers of soap, coffee, and automobiles. Investment analysis is largely in a pre-theoretic stage of development. Security analysis, narrowly defined, consists chiefly of naïve extrapolations from ratios based on accounting data." Boness wanted to replace this unsophisticated, nonrigorous approach with a sound, well-founded mathematical theory.

Boness described a model for the price of options. He then compared the prices predicted by the model to the actual prices of options traded in the New York market. As readers could glean from his graph, the fit was quite good, if not perfect. But in developing the model, Boness had made a couple of assumptions that were not warranted. "For convenience and in default of better information, all stocks on which options are traded are defined to be of the same risk class." As assumptions go, this was a rather weak one. We would not be in very good shape if convenience and the lack of information were the guiding lights of science.

There was more. Boness also posited that investors are indifferent to risk. If throwing all stocks into one basket had not been enough, now all investors were thrown into the same basket as well. Worse, it was the wrong basket because the vast majority of people are not indifferent to risk. To the contrary, they are risk-averse. But most scientific advances begin with small steps and, given the state of the theory of option pricing, Boness, by incorporating the time value of money, set the stage for those who were to follow. And anyway, the assumption about the investors' risk attitudes was false.

All attempts so far, including those of Sheen Kassouf, Edward Thorp, Case Sprenkle, James Boness, Robert Merton, and Paul Samuelson had their shortcomings, from Bachelier's possibility of negative stock prices to many of the later authors' incorporation of investors' risk attitudes. But amazingly, their equations, though not totally correct, carried the seeds of the eventual solution. They contained heavy clues to what would eventually turn out to be the correct options pricing formula. But they had hit upon their alleged solutions in an incorrect manner. Unable to come up with the proper discount rate, they assumed that in order to buy options, risk-averse investors would demand a premium over and above the risk-free interest rate. Hence their formulas relied on an unknown interest rate that included compensation for the risk associated with the stock. As it turned out later, however, the risk premium is already contained in the price of the stock; consequently it was wrong to add an additional premium for the option. Black, Scholes, and Merton had to pool their efforts to derive the options pricing formula in a correct way, as described below.

––––––––––

Sometime in 1968 or 1969, at roughly the same time that Samuelson and Robert Merton were working on the subject, Fischer Black was starting to get interested in a formula for the value of a warrant. His starting point was

CAPM, the capital asset pricing model. As explained above, CAPM states the following: first, an asset's risk is composed of a systematic component that reacts to the market in general, as well as a component that is intrinsic to the specific asset. The latter can be eliminated by diversifying the portfolio. Second, assets that carry more risk must have higher expected returns than less risky ones. However, investors are only compensated for the systematic component of the risk; they are not remunerated for the risk that they could diversify away.

To reduce the model's complexity, Black made a few simplifying assumptions. He postulated that trading costs are zero—that there are no brokerage fees. He also assumed that investors can borrow and lend money at the same interest rate. Both assumptions are unrealistic, of course, since brokers charge fees for their services and banks demand higher interest rates from borrowers than they pay to lenders. Finally, Black also assumed that the stock's volatility is constant.

He then went on to derive an equation which said simply that the expected return on a warrant should depend on its risk in the same way that a common stock's expected return depends on its risk. Applying CAPM to every moment in a warrant's life, for every possible stock price and warrant value, he thus arrived at a differential equation whose solution would give the true, intrinsic value of the warrant. (See the Pedestrian's Guide at the end of the book for an intuitive, nonrigorous synopsis of how the equation can be derived.)

To his utter surprise, Black noted that the equation that he had derived did not contain the stock's expected return as one of the variables. This implied that the warrant's price is not affected by the stock's expected return. It was quite the opposite of what had hitherto been thought. Until then, everybody had been certain that a warrant's price must reflect the stock's expected return. And another surprise was in store. The separation of risk into the two components, the stock's intrinsic risk and the risk that can be diversified away, did not appear in the differential equation either. Hence CAPM's dichotomy did not express itself in the price of the warrant.

There was a catch, however, and it was serious: Black could not solve the equation. The holy grail was in sight but he could not reach it. In vain he spent days looking for an answer. It was like embarking on an expedition to Mount Everest and then getting stuck just below the peak. Worse still, it was his own fault for not having been as diligent a student as he should have been. "I have a Ph.D. in applied mathematics," he wrote later, "but had never spent much time on differential equations, so I didn't know the standard methods used to solve problems like that." Coming from a Harvard graduate, the admission is quite amazing, since differential equations are the bread and butter of applied mathematics. And he could have also been spared frustration had he at least been more attentive as an undergraduate. But that had not been the case either. "I have a B.A. in physics, but I didn't recognize the equation as a version of the 'heat equation.'" As was pointed out in an earlier chapter, the significance of the heat equation for the pricing of options had already been noted by Bachelier in 1900. Had Black been aware of the equation's identity, he would have easily succeeded in solving it, since the heat equation had a well-known solution.

Alas, it was not to be. Black was stuck. Frustrated, he dated his notes, placed them in a folder, and put the problem aside.[2] Thus for the time being at least, the advance that Black had managed to chalk up remained hidden in a filing cabinet.

But not for long. Black's company, Associates in Finance, continued its consulting activities and sometime in the fall of 1969, the options problem came up in a conversation with his sole associate, Myron Scholes. Black pulled his file and the two began working on what would eventually become one of the most famous equations in finance and make "Black and Scholes" a standard term used by business students and finance professionals all over the world.

Progress was rapid. They threw in a few more assumptions, like the absence of taxes, reinvestment of all dividends, and continuity of stock prices. (The latter assumption implies that stock prices make no jumps. With a share price moving from, say, $10 to $11, it would have to visit all prices in between: $10.01, $10.02 . . .) While the reinvestment of dividends is a very reasonable assumption, absence of taxes and the continuity of stock prices are not. But the assumptions were postulated only to simplify the theoretical derivation. By relaxing the assumptions, the model could always be rendered more realistic later on. The basic idea on which Black and Scholes built their model was brilliant—create a hedged portfolio and see where that led them.

Hedging is a method to eliminate riskiness. To "hedge your bets" means doing one thing while also doing the opposite, such that, on average, you neither gain nor lose. At a roulette table, for example, you could bet $1 on red and another on black. After the wheel has been spun, you will certainly lose one of your dollars but, at the same time, you will receive $2 for the winning bet. No matter where the ball falls, you end up with exactly what you started with: $2. Hence you eliminated risk.

The roulette example is not a perfect hedge, of course, because every once in a while the ball falls on the green zero, in which case the house keeps both of your dollars. But as economists are wont to do, let's make an assumption: the roulette wheel has no zero. It is just like Black and Scholes's assumption that there are no transaction costs. In this case, placing one chip each on red and black is a perfect hedge. But if you end up with the same amount that you started with, what sense does it make to play?

It doesn't, unless something is mispriced!

Let's say that the casino pays you $2 plus a penny if the ball falls on the color on which you put your dollar. In this case, you could play over and over again, gaining a penny every time. If that is too slow for you, why not borrow $10 million, bet half the sum on red and the other half on black. After a single roll, *les jeux sont faits,* the house will pay you $10,050,000. You would be left with a large bundle even after you paid back the interest on your loan. Obviously there was a mispricing: by selling roulette chips for an even dollar, while paying out $2 and a penny for successful bets, the house had undervalued its chips.

I described a hedge at the roulette table because this is exactly how Black and Scholes proceeded. Let's recall how options work. A call option is a contract that gives the buyer the right to demand delivery of one share at the price of, say, $100 at a specified date in the future. If the share trades at $110 on the specified day, the buyer will exercise his right to receive the share because he can immediately sell it on the market and make a $10 profit. If the share trades at a price below $100, he will let the option expire without doing anything.

We see that whenever the share price rises, the potential profit to the holder of the option also increases. Hence investors will be willing to pay more for the option and its value rises. Let's assume for now that the share price is far above the striking price. In traders' lingo it is "deep in the money": it is nearly certain that the option will be exercised and the holder will make a profit. In this case a further $1 increase in the price of the share will increase the option holder's profit, and hence the value of the option, by a dollar.

Before I explain how to hedge a financial portfolio, we need to review what it means to sell a stock short. In exchange for a certain sum, the seller promises to deliver the share to the buyer at some specified date in the future. On that day, he will have to purchase the share on the market in order to fulfill his obligation. Let's say the short seller received $100 when the short sale was effected. If, on the settlement date, the price of the share is above $100, the seller suffers a loss, since he will have to buy it at a price higher than the one he received. If the price is below $100, he will make a profit.

To construct a hedged portfolio if an option is deep in the money, purchase one option and short sell one share. If the share rises by a certain amount, the investor loses this amount due to his short sale. But this loss is exactly offset by the other part of the portfolio, since the value of an option rises when the share price goes up. On the other hand, if the share price decreases by a certain amount, the short sale engenders a profit that is offset by the loss that the option engenders. Whatever happens to the share price, the investor ends up with the exact same amount he started with. But how many options are required to offset the risk of holding one share short? In the above example, the option's value increased by exactly the same amount that the share's value increased. Hence it took one option to offset the risk of one share. The so-called hedging ratio, defined as the number of options required to hedge a portfolio that contains one share held short, is one.

Different versions of hedged portfolios can be created by alternatively buying or selling puts and calls, by reversing long and short positions of either the stock or the option. It gets a bit confusing but the upshot is this:

by going long on one financial instrument and short on the other, a judicious investor can eliminate all risk from his portfolio.

What if the option is not deep in the money? In this case uncertainty abounds and we have to look at the situation afresh. Let's say that the exercise price is $100 and the share price currently hovers around $101. The investor has a chance of making a profit, but there is a real possibility that the share price drops below $100 and he will have to let the option expire. Hence the option's value is still unknown. Can we at least determine by how much the option's value rises if the share price increases to $102? Because of the uncertainty, it will not rise by exactly $1, that's for sure. But more than that, we do not know for the time being. Hence we also do not know how many options are needed in order to construct a hedged portfolio. For argument's sake, let us say that the option's value rises by 50 cents for every $1 rise in the share price. In this case, the hedged portfolio would have to be composed of two options for every share that is being sold short. If the share price increases by a dollar, the two options increase by 50 cents each, thus balancing out the loss on the short sale of the share. In this case, the hedging ratio is two. But we don't yet know by how much the option's value rises or falls in response to a change in the price of the underlying share. This, and the actual value of the option, is what Black and Scholes wanted to find out.

So, in order to create a riskless portfolio, what should the ratio of options be to the number of shares sold short? And, above all, how does information about the hedging ratio help determine the true value of an option? As Black and Scholes wrestled with these questions, they had another idea.

As noted above, mispricings are a reason to join a game of chance. But there is another sound economic motive to embark on such ventures: the question of time. In principle, investors should not spend days playing around with hedged bets or portfolios, if all they can hope for at the end is the sum that they started with. In such a case, they could have just as well kept the money in a savings account and received the interest on it. To induce people to put money in hedged portfolios, the investments must return something. Given the Brownian motion of the stock prices, the pricing of options must therefore be such that there is some payoff over time.

This is where Black and Scholes applied the ax. A hedged portfolio is totally without risk. No matter what the market does, according to CAPM such a portfolio should return neither more nor less to the investor than an equally riskless deposit in a savings account or an investment in U.S. Treasury bills.* This insight enabled Black and Scholes to write down the crucial equation. With the unknown hedge ratio as a variable, the left-hand side expressed how the value of the hedged portfolio grows over time. On the right-hand side they put the riskless rate of interest. (See the Pedestrian's Guide at the end of the book.)

So now they had arrived at a differential equation. But this was only a partial success. It was where Black had gotten stuck a few months previously. A differential equation expresses how things change over time. In this case it is the change in the value of the hedged portfolio that is being expressed. The solution to the differential equation then gives the actual value of the portfolio, with the hedging ratio and the correct price of the option as by-products.

If Black had not had the wherewithal to solve the differential equation, Scholes was not much better at it either. But together they progressed. After they inspected the equation in many different ways, they made a guess at the solution, albeit an educated one, and then checked whether it fit. It did!

Black and Scholes had found the pricing formula for options. It looked formidable:

$$C = SN(d_1) - Xe^{-rT}N(d_2),$$

where

$$d_1 = \frac{Ln\left(\frac{S}{X}\right) + \left(r + \frac{1}{2}\sigma^2\right)T}{\sigma\sqrt{T}}$$

and

$$d_2 = d_1 - \sigma\sqrt{T}$$

---

* See footnote on page xiv.

Let me explain the various symbols, the easy ones first. C stands for the price of the option or warrant, which is what we want to compute. S is the current share price, and X is the exercise price. T is the number of periods—days, weeks, months, or years—remaining until the exercise date; r is the risk-free rate of interest per period. (The period must be measured in the same units as T.)

Now we get to the more difficult symbols and expressions. Both terms on the right-hand side contain $N(d)$. N stands for the normal distribution. If you draw a number randomly out of a standard normal distribution, $N(d)$ denotes the probability that the drawn number will be less than d. Being a probability, $N(d)$ obviously is a number between zero and one. Understanding d_1 and d_2 is more difficult still. Ln is the natural logarithm and σ is the stock's volatility—the standard deviation of its returns. The square of the volatility, σ_2, is the variance. Sometimes T, the time to maturity, appears under a square root sign. This, the reader may recall, is the "square root of time" law, a consequence of the Brownian motion of stock prices, as Bachelier, Einstein, Smoluchowski, and Langevin had found out a long time ago.

So the first term on the right-hand side, $SN(d_1)$, denotes a fraction of the stock price. If S is, say, \$120 and $N(d_1)$ is 0.5, then $SN(d_1)$ equals \$60. What does this number mean? It turns out the $1/N(d_1)$ is the hedge ratio that is required to make the portfolio riskless. Hence for every share that the investor holds short, he should hold two options (1 divided by 0.5). If the share price rises by a dollar, the options will each rise by 50 cents, thus leaving the value of the portfolio unchanged. Or, viewed from another angle, for every \$60 held in shares, the investor should hold one option short (i.e., for one \$120 share he should hold two options short). Then his portfolio will be riskless.

Now let's take a close look at the second term on the right-hand side of the equation. It is composed of a few subparts. First, let's examine $-X$. If, on the exercise date, the price of the share is higher than the exercise price, the investor will exercise his option and pay the agreed amount to the seller of the option. Since he pays this amount, the X is preceded by a minus sign. Since the payment will be made only T periods hence, it needs to be discounted. This is expressed by multiplying $-X$ with the so-called discount factor e^{-rT}. (e is the exponential function, the inverse of the natural logarithm, which—for reasons we will not go into—is used to compute the present value of future payments. r is the risk-free interest rate and T, of course, is the time remaining until the exercise date.) Remember, however, that it is by no means certain that the option will be exercised. There is only a probability that this will happen, which is expressed by multiplying $-Xe^{-rT}$ with the fraction $N(d_2)$. Whew!

So the formula says that the options price is equal to a certain fraction of the stock price, minus a fraction of the discounted exercise price. If the current stock price is "deep in the money"—if S is much larger than X—the fractions are close to one. Then the call option is approximately the difference between the stock's current price and the present discounted value of the exercise price. If, on the other hand, the current stock price is "deep out of the money," S is much lower than X, making the value of the call option close to zero.

Altogether, the option's value depends on five variables: the price of the underlying stock, the option's exercise price, the time to maturity, the risk-free interest rate, and the variability of the stock's price movements. One variable is notably absent. As Black had noted to his complete surprise when he first developed the differential equation, the price of the option does not depend on the expected return of the underlying share. This was a complete reversal of what had previously been thought. Until Black and Scholes came up with the solution to the differential equation, most researchers were convinced that the value of an option must somehow depend on the share's expected return. This erroneous belief is what had led earlier economists astray.

Four of the five variables are known or can easily be determined: the option's exercise price and time to expiration are specified in the contract. The price of the underlying share can be checked on the stock exchange ticker. The risk-free rate of interest is usually taken to be the return on U.S. Treasury bills with a maturity date similar to the one of the option.* Only the fifth

---

* See see footnote on page xiv.

variable, the stock's variance, cannot be observed directly. Generally, past volatility is considered a good indicator for the share's present and future variance. But predicting the future based on the past has its shortcomings.

An options trader needs to know how the price of the option reacts to changes in each of the five variables. Mathematically, this can be determined by taking the first derivative of the expression for the options price with respect to the variables and checking whether the result is positive or negative. (In what follows, we concentrate mainly on call options.) As the price of the stock rises, so does the price of a call option. After all, the higher the stock, price the better the chances that the investor will make a profit. On the other hand, the higher the exercise price, the lower is the option's value. This is so not only because the probability that the option won't be exercised is higher, but also because the investor will have to pay more to obtain the share when he eventually does decide to exercise the option. The higher an option's price, the longer the time to expiration. One reason for this is that the discounted value of the final payment is lower if the exercise date is a long way off. Expressed differently, as time progresses, the option's price decays. A high rate of interest also implies a high options price. Again, one of the reasons is that a high interest rate means that the present value of the final payment, the discounted exercise price, is low, thus raising the value of the option.

Finally, the higher the variance, the higher the value of the option. Understanding this is a bit trickier. A large variability in the stock's price means that there is a chance of large positive changes. This is good for the holder of a call option. On the other hand, the probability of large negative changes is also greater. This is bad for the investor. Drops below the exercise price can be ignored, since the option won't be exercised. Hence the positive changes outweigh the negative ones. That is the reason why large variability raises the call option's value.

To denote the sensitivity of the options price to the variables, Black, Scholes, and those who followed them used letters of the Greek alphabet. The "Greeks" indicate not just in which direction the option price will move in response to fluctuations in a variable but also by how much.[3] For example, the delta gives the sensitivity of the price of the option to changes in the price of the underlying stock. A delta value of 0.5 means that the call option's price increases by 50 cents for every dollar increase in the share price. Options that are very deep in the money usually have a delta of close to 1.0, meaning that a dollar increase in the share price will increase the price of the call option by nearly a dollar. An option's gamma indicates how its delta changes. It is an indication of the change of the change.[4] The deeper the option is in or out of the money, the smaller the gamma because with its value close to 1.0 the change must slow down. Theta measures how the option's price changes as time progresses toward the expiration date; rho does the same for the interest rate. Finally, vega (which sounds like a Greek letter but is not) indicates how the option price reacts to a change in the underlying stock's volatility. There are also more exotic Greeks like the zomma (again, not a Greek letter), which measures the rate of change of gamma with respect to changes in volatility.

Black and Scholes's derivation of the option pricing equation is based on the postulate that the portfolio be risk-free (i.e., that it be hedged). With stock prices continuously changing—in fact, it is a characteristic of Brownian motion that the stock price changes at every instant—the correct hedging ratio changes continuously.[5] In order to keep the portfolio risk-free at all times, investors would need to adapt the composition of their portfolio by the minute or even faster. They would continuously buy and sell shares and options in order to keep a balanced portfolio of long and short positions.

This is where the option's delta comes in handy. It is a crucial parameter. Since it indicates how the value of the option changes with an alteration in the price of the underlying share, delta determines the correct ratio of call options to the number of shares held short. Even a slight change in the price of the share would require the investor to buy or sell a certain number of options. Theoretically, this is no problem. Since Black and Scholes assumed that there are no trading costs, the investor can continuously update his position. But in practice, every purchase and every sale involves brokerage fees. To keep the portfolio totally risk-free at all times would involve an immense number of trades, and soon the costs would accumulate to more than any profits the investor could hope to gain. Jules Regnault had said so more than a century earlier. This is where the gamma comes in handy. Recall that it is a measure of how the delta changes, thus indicating how important it is to adapt the position. A low value for gamma means that the delta, and hence the hedging ratio, is less sensitive to a small price change and there is less need to adapt the position as often.

————

Now that they had solved the differential equation, Black and Scholes were ready to present their findings to the world . . . or at least to their clients. They got their chance at a conference that Scholes organized at the end of July 1970. It was the second Wells Fargo conference on capital market theory. The two presented their findings in a morning session. Their paper was titled "A Theoretical Valuation Formula for Options, Warrants, and Other Securities."

Unfortunately, one person who would have greatly appreciated the significance of the talk was not present. It was Robert Merton. The freshly minted Ph.D., whom Scholes had interviewed for a position at MIT's Sloan School a few months earlier, was to give a talk of his own at the conference titled "A Dynamic General Equilibrium Model of the Asset Market and Its Application to the Pricing of the Capital Structure of the Firm." But on this morning Merton overslept. So it was only after he gave his own presentation in the afternoon that the three men realized that they had been working on the same subject. This is quite amazing, given their proximity at MIT. Merton had cooperated with Scholes on various subjects, but options pricing was not one of them. Actually, as academics always do, Black and Scholes had made a literature search to see how close other researchers were to their work and had come across the earlier paper by Merton and Samuelson. They may have, or should have, suspected that Merton was continuing his work along these lines. But since they were desperately trying to get their own derivation in order, they did not tell anybody about it.

Black admitted as much in his reminiscences twenty years later. "Neither of us told the other. We were both working on papers about the formula, so there was a mixture of rivalry and cooperation." And Scholes attests that "Fischer and I wanted to progress, on our own, as far as we could prior to the conference."

Following the afternoon session, Merton asked Scholes to explain what he had missed while he slept and the two discussed their work. Merton was not convinced by Black and Scholes's argument. In particular, he did not believe that the portfolio could be made completely riskless by hedging. You may recall that in their model the hedging ratio needs to be adjusted as time progresses and whenever the price of the underlying share changes. Merton thought that as the interval between adjustments of the hedging ratio becomes ever smaller, in effect moving toward continuous trading, some risk would remain in the portfolio.

In the weeks that followed, Merton was busy with his own version of an options pricing formula. He tried a different tack—the creation a portfolio that would replicate the payoffs of an option. The cost of purchasing such a portfolio would then have to be equal to the price of the option. After some hard work, Merton found that by buying a certain number of shares and borrowing a certain amount of money at the risk-free interest rate, he could exactly simulate the investment in an option. In order to hedge away all random fluctuations that arise due to the Brownian motion of stock prices, Merton's methodology required the continuous adjustment of the portfolio. This is where the techniques developed by Kiyoshi Itō became indispensable. Making copious use of Itō's stochastic calculus, Merton was able to create a portfolio, at least on paper, that would be riskless at every moment in time. True, the amounts would have to be adjusted at every moment throughout the life of the option but since he, like Black and Scholes before him, also postulated that trades did not involve any costs, this presented no problem.

The upshot of all this was that the price of the replicating portfolio, which could now be calculated, would have to be equal to the price of the option. It had taken several weeks, but now Merton was finally there; a few more mathematical manipulations and he was in possession of the sought-after equation. When he looked at it, he was thunderstruck: the expression he had obtained was identical to the equation that Black and Scholes had found.

It was Saturday but Merton did not hesitate. He telephoned Scholes to tell him the news: he and Black had been right; his own model gave the same result as the one they had developed. By different routes, both Black and Scholes and he had found the correct price for an option.

Having presented the options price formula to a select group of partic-ipants at the Wells Fargo conference, the time had come to publish it in a respected scientific journal. But Black and Scholes quickly found out that this was no simple undertaking. A first submission to the *Journal of Political Economy*, edited at the University of Chicago, was rejected. Adding insult to injury, the paper was returned without even being reviewed by an expert in the field. The editor simply wrote that the subject matter was too spe-cialized for them. This was quite amazing, since seven years earlier Boness's paper had appeared in the *JPE* on exactly the same subject. An inquiry sub-mitted to the *Review of Economics and Statistics*, edited at Harvard, fared no better. Black ascribed the rejections to their lowly standing. After all, Scholes was just an assistant professor, and Black was not even that. "I sus-pected that one reason these journals didn't take the paper seriously was my non-academic return address," he wrote in an article titled "How We Came Up with the Option Formula" in the *Journal of Portfolio Management* in 1989.[6]

That would change in May 1971. Black was invited to give a talk at the University of Chicago. Just six days after his "maiden lecture," the dean called with an offer of a Ford Foundation Visiting Professorship of Finance for the coming academic year. This was the occasion Black had been wait-ing for; it was his chance to make up for the lack of academic credentials. He accepted, closed his office near Boston, sold his house, and moved to Chicago with his family. In September he assumed his new duties.

With his inquisitive nature and sharp mind, Black fit right into the open, intellectually challenging atmosphere at Chicago. He enjoyed the intellec-tual sparring with colleagues, especially with the Nobel Prize laureate Milton Friedman, founder of the so-called Chicago School of economics.[7] It is no surprise therefore, that the university's Graduate School of Business upgraded him to the status of full professor only one year later. In 1973 his consulting associate and academic partner Myron Scholes joined him on the faculty. Their strong belief in perfect markets and in the rational behavior of investors stood them in good stead at Chicago.[8]

From then on, Black's return address was more in line with what editors of prestigious journals expected. Moreover, he and Scholes now had colleagues who could aid them in their struggle to have their paper published. Merton Miller and Eugene Fama, comrades in arms at the University of Chicago, started to take an interest in the paper and began to intervene. First, they made suggestions on how to render the paper more palatable to journal editors. Second, they suggested to the powers that be at the *Journal of Political Economy* that this paper was worthy of another, more serious look. The editors relented. They had the paper reviewed and accepted it, subject to some revisions. In the May-June 1973 issue "The Pricing of Options and Corporate Liabilities" was finally published. It would become one of the most important papers ever in economics. According to a study undertaken in 2006, twenty-three years after its publication, it had garnered 2,589 citations, making it the sixth most cited paper in all of economics.

Ironically, while Black and Scholes were still revising their paper for the *Journal of Political Economy,* another paper of theirs, in which they empirically tested their as yet unpublished formula, had been accepted by the unrefereed annual sessions volume of the *Journal of Finance.* The paper closed by acknowledging Merton's work: "Robert C. Merton has developed the same formula as ours, starting from somewhat different assumptions. The knowledge that his formula agrees with ours gives us greater confidence that we haven't made any substantive errors along the way."

Merton had an easier time with his paper: he was asked to publish it. MIT professor of management Paul MacAvoy, the founding editor of the *Bell Journal of Economics and Management Science,* asked Merton in 1971 whether he would agree to publish his derivation of the options formula in his newly founded journal. Of course Merton, who had only recently obtained his

Ph.D. and was still trying to make a name for himself, agreed. But he wanted to be fair to his senior colleagues, whom he acknowledges in the abstract by referring to "the seminal Black-Scholes theory of option pricing." He stipulated that his paper would appear only after the one by Black and Scholes had been published. So his "Theory of Rational Option Pricing" lay in wait for two years before it eventually appeared, also in 1973.

Much had happened in the decade since Boness lamented the dearth of theoretical progress in the field. Thus Merton's paper could open with the words "the theory of warrant and option pricing has been studied extensively in both the academic and trade literature." He went on to wonder why that was, since options were a specialized and relatively unimportant security at the time. Merton attributed the growing interest to the fact that options were a particularly simple contingent-claim asset and that the theory could be extended to a general theory of contingent-claim pricing.

options
derivatives
oh
my!

SIXTEEN

The Higher
They Climb...

FISCHER BLACK AND MYRON SCHOLES JOINED THE UNIVER-
sity of Chicago at a judicious time. On April 26, 1973, just one
month before Black and Scholes's article was published, the
Chicago Board Options Exchange (CBOE) opened its doors.

With a myriad of possible exercise prices, exercise dates, and option
prices, it would be impossible to have a market for stock options without
some kind of standardization. The CBOE was the first exchange in the
United States to list standardized options for trading. Its founders decided
that the exercise dates would be the Saturdays immediately following the
third Fridays of the four settlement months. The first two settlement
months are the current month and the next, if the current month's settlement date has not yet been reached. Otherwise they are the next month and
the one after that. In addition there are two more settlement months within

the next nine months, depending on when the option was first issued. (There are longer-term option contracts that can have expiration dates up to three years after the date of the listing.) Exercise prices move in increments of $2.50 when the strike price is between $5 and $25, $5 for strike prices between $25 and $200, and $10 when the strike price is above that. The prices of options move in intervals of nickels for options trading below $3, and in dimes for options trading above.

In addition to being standardized, a market must also be trustworthy. If investors cannot be sure that their orders will be executed and that they will receive their dues on the exercise date, they won't be willing to trade. Hence, the CBOE regulates, monitors, oversees, enforces, arbitrates, and, if the need arises, brings legal action against offenders. To provide transparency, the CBOE also makes available immediate and public reporting of all transactions.

But a market will not develop even if trading is standardized and the market place is trusted, if there are no takers. This is where the so-called market makers jump in. They are members of the CBOE who provide liquidity in the marketplace by risking their own capital. In the absence of buy or sell orders from the public, they make bids and offers for their own accounts. Market makers are the backbone of the CBOE's trading system.

Over the years, all these efforts paid off to make for a very successful operation. On CBOE's first day of operation in 1973, 911 contracts were traded. By the middle of 2009, the daily volume on the CBOE had increased to nearly 5 million contracts a day, with occasional spikes to 10 million. And while sixteen underlying stocks were traded on opening day, a quarter of a century later there were more than two thousand. A seat on the CBOE, originally priced at $10,000, was sold for over $3 million in mid-2008. (By mid-2009, after the financial crisis, the price had dropped to a little below $2 million, and was back to $2.5 million in mid-2011.)

The Black-Scholes formula became a cornerstone of options trading. Two years after the CBOE was founded, the board officially adopted the equation to price options. Then the manufacturers of computing equipment saw a new market. Texas Instruments was the first to introduce handheld calculators that could compute options prices according to the Black-Scholes equation. Hewlett Packard quickly followed with a comparable model of its own, the HP-41.

The TI Programmable 59 sold for $299.95 and allowed 960 program steps. Furthermore, programs with up to 5,000 program steps could be stored either on magnetic cards or on interchangeable modules. Texas Instruments' Securities Analysis module contained program SA-07, which computed options prices according to Black and Scholes.

Now it was only a hop, skip, and jump to compute hedge ratios and option values. After selecting the program and initializing, the user entered the stock's current price, its exercise price, its volatility, the riskless rate of interest, today's date, the expiration date, and the dates and amounts of dividends—if there were any. A final press on the "R/S" key computed the hedge ratio, a press on the "E" key gave the option's value. As one of the discoverers of the options pricing formula, Merton thought he deserved to participate in Texas Instruments' success. But he was quickly sent packing. "When I asked them for royalties, they replied that our work was in the public domain; when I asked, at least, for a calculator, they suggested that I buy one. I never did."

Handheld calculators were the actual catalysts for the adoption of the Black-Scholes formula by professional traders. At a time when computers were still monstrous mainframes, these handhelds permitted sophisticated computations from the trading floor. In an interview with Options News Network TV in 2008, Bill Brodsky, the chairman and CEO of the Chicago Board Options Exchange, recalled that "when options started [to be traded on the CBOE] they didn't have those equations. Black-Scholes became part of the option world only after Texas Instruments created a calculator that could do instantaneous calculations of complex formulae, and it was then that the Black-Scholes formula became embraced in the option world. But it took a couple of years, and so there were people who were trading options literally from the seat of their pants before Black-Scholes."

By the mid-1970s, there was a well-developed market for options, investors possessed a mathematical formula that gave the options' correct values, traders had calculators that computed these values, and market makers used these values to determine prices. So, to paraphrase Voltaire, was all for the best in the best of all possible worlds?

Economists and financial theorists had their doubts. After all, in the real world, prices are determined by supply and demand. Do option prices on the market really conform to the formula? Or is this theory a chimera, albeit one with elegant theoretical underpinnings? Academics clamored for an empirical verification of the equation. Only if a test were performed and the results turned out as predicted—if the prices at which options are actually traded conform to the prices computed by the formula—could they be sure that the theory was correct and not just a nice toy model.

The earliest large-scale empirical investigation of options prices was performed by Dan Galai, a young Israeli Ph.D. student at the University of Chicago. With Merton Miller and Myron Scholes on his doctoral committee and Fischer Black as an adviser, he was well placed to perform the task. For the first time, real-life option prices would be subjected to rigorous testing. The CBOE supplied the data on punched cards—daily prices for each option traded on the CBOE during its first seven months of operation (April 26 to November 30, 1973). To verify that the cards contained no punching errors, the data was checked against the closing prices reported daily in the *Wall Street Journal*.

Galai was going to verify whether the CBOE's data conformed to what the Black-Scholes equation predicted. The test consisted in checking whether the prices of the options on the CBOE allowed an investor to make a profit. Recall that Black, Scholes, and Merton derived the prices of options by creating hedged portfolios of stocks and options. By buying one security and shorting the other, the portfolio is totally riskless. Hence, over time, it should return exactly as much as the risk-free rate of interest. (The latter is defined as the rate at which one can borrow or lend money at the bank.) The correct value of the option can then be deduced by calculating—via the notorious differential equation—the price at which the portfolio returns exactly the riskless rate of interest.

If the option were priced incorrectly, investors could make an easy profit without assuming any risk. Let's say that the price of the option is too low.

The investor can then borrow money at the riskless rate of interest and invest it in a hedged portfolio by buying options and going short on stock. Since the price of the option is lower than it should be, this portfolio will return more than the riskless rate of interest. Thus, without taking on any risk, the investor will receive more than the interest that he would have to pay the bank. (If the price of the option is too high, the investor will make money by going short on the option and buying stock, again borrowing money from the bank.)

On the other hand, if options were priced correctly, then, according to the model of Black and Scholes and Merton, no profits were possible. This is where Galai applied the lever. If investors could have made money during the 152 trading days between April 26 and November 30, 1973, by investing in hedged portfolios, then something would be wrong.

Something would be wrong, but what exactly? Galai's work would test two hypotheses simultaneously: first, that the Black-Scholes model is valid and, second, that the market at the CBOE is efficient. If the test went awry, it could be that either hypothesis was incorrect, but the test would not have revealed which. Luckily, it would not come to that.

From the data he received, Galai built hedged portfolios according to the Black-Scholes model. He made the assumptions that both the riskless interest rate and the stocks' volatility stay constant during the option's lifetime. (With the chairman of the Fed affecting the interest rate on a monthly basis, at least the first assumption was not quite warranted. The second was even more problematic given the difficulties of estimating volatility.) The value of the portfolio was adjusted for dividend payments and the hedging ratio was adjusted daily according to the Black-Scholes model.

The findings were as hoped for, even if they may have seemed slightly ambiguous at first. "The market did not seem perfectly efficient to market makers," Galai concluded. "The ex ante returns, while usually statistically insignificant, show a strong tendency to be positive." That would seem a denunciation of the whole theory, since the existence of profit possibilities indicates either that the model predicts incorrect option prices or that the market is not efficient. But note that Galai referred only to market makers. These are the members of the CBOE who buy and sell options for their own account. Since they do their own trading, they incur no brokerage fees. The saving is especially significant since many transactions are required in order to maintain the correct hedge ratio.

What happens if the brokerage fees that a regular investor incurs are taken into account? Galai redid his analysis with transaction costs of 1 percent. Now all profits were effectively washed out. As is customary in academic papers, he stated his conclusion with restraint. "Inclusion of transaction costs might eliminate this positive drift. In any case, it does not seem that a nonmember of the CBOE can expect to achieve 'above-normal' profits consistently." The upshot was that regular investors cannot expect to make profits. Hence we can conclude both that the model holds and that the market is efficient. It was the first vindication of the theory of Black, Scholes, and Merton. Galai's thesis was accepted in 1975. On his way back to the Hebrew University of Jerusalem, where he would be named lecturer and eventually professor, the freshly minted Ph.D. may have felt a little like Sir Arthur Eddington on his return from the island of Principe. The English astronomer had led an expedition to this island off the coast of West Africa in order to measure the deviation of a star's rays around the sun during a solar eclipse in 1919. It was the crucial evidence that was needed to lend credence to Einstein's theretofore speculative theory of relativity.

After moving from Chicago to Boston in 1975, Fischer Black remained at MIT for nearly ten years. In 1984 he decided to put his money where his mouth is or, more aptly, to put his mouth where the money is: he joined Goldman Sachs.[1] Actually it was Robert Merton who got Black the job. He was doing consulting work for Goldman Sachs and suggested to his client that they hire someone full-time who could incorporate the newfangled academic theories into their business. Initially Merton thought it would be an opportunity to place one of MIT's brilliant graduate students, but when he told Black about the possibility, Black became intrigued and offered himself.

Black became the first so-called "quant," a quantitative analyst who develops mathematical models for financial investments, that Goldman Sachs ever hired. In fact, he may have been the first quant on Wall Street. As such he was a special sort of employee. With one day a week and the weekends reserved for his own projects, Black continued doing research as he had done at the University of Chicago or at MIT, albeit with a much higher salary. Soon many colleagues questioned his usefulness to the firm. But Robert Rubin, an important partner (and later secretary of the Treasury in the Clinton administration), quieted the critics. Black's job was not to create fancy new financial products, meet with clients, or make deals. His mission was to infuse new ideas into the firm. Black soon proved himself worthy of Rubin's trust. Ever the outsider, he made a name for himself as a theoretician with a bent for real-life financial applications. (Within academia he had been known as a practitioner with a bent for theoretical pursuits.) And having Fischer Black on the roster did not hurt the firm's reputation with clients either. Within two years, Black became a member of the most hallowed group of people within Goldman Sachs: he was elected a partner in the firm.

Myron Scholes remained on the Stanford faculty from 1981 until he retired from teaching fifteen years later. In 1990 he also caught the bug and decided to get involved more directly with the financial markets . . . and make a little money on the side. He joined Salomon Brothers.[2]

Scholes joined the investment bank as a special consultant, eventually becoming a managing director and co-head of its derivatives group. In contrast to Black at Goldman Sachs, Scholes was not hired to perform cerebral activities. "He'll be seeing clients and working with clients, and he will not be doing research," the other co-head of derivatives told the *New York Times*. Scholes put it this way: "I am going to be eclectic for a while." He retained his professorship at Stanford, however, teaching the winter quarter and working for Salomon the remaining nine months. With a six-digit base pay and an annual bonus that could top the salary, it was a very advantageous arrangement.[3]

At Salomon Brothers, Scholes reconnected with Robert Merton, who had already been there for two years. In 1988 Eric Rosenfeld, one of Merton's former students at MIT, had paid him a visit, together with the company president, Thomas Strauss. They offered him the position of special consultant to the Office of the Chairman. It was an offer Merton could not refuse. In his previous consulting engagements, among them with Goldman Sachs, he had mainly built models and designed financial products. The position that Salomon was offering would give him the opportunity to advise the chief executive officer on business matters and influence the firm's direction. With Salomon Brothers being one of the really big players on the financial scene, the new position would quite possibly put him in a position to bring about institutional change in the global financial system.

For a few years Black, Scholes, and Merton had fun, pursuing their research and making heaps of money. But then tragedy struck. In February 1995, Black was diagnosed with throat cancer. Surgery seemed successful and he started to recover. But the cancer returned and in the summer it became clear that Black did not have much longer to live. His friends and colleagues at MIT and at Goldman Sachs decided to arrange a collective tribute as a final sign of appreciation. They informed him of their plans in a letter dated July 14.

"Dear Fischer," they wrote. "Your many friends, colleagues, and admirers want to honor you and to celebrate your seminal contributions to finance and economic science. They have decided to do so in three parts: a chair, a prize and a collection of your scientific papers." The Fischer Black Visiting Professorship of Financial Economics would be established at MIT's Sloan School of Management. The decision to endow a visiting professorship and not a chair was meant to avoid intellectual inbreeding and stagnation. Inviting a new visitor every time would help develop and exchange new ideas and perspectives.

The Fischer Black Prize would be awarded every two years by the American Finance Association "for a body of work that best exemplifies the Fischer Black hallmark of developing original theoretical research concepts that have a direct and significant impact on finance practice." In a third expression of appreciation the friends decided to directly honor Black's scientific work. The publication of his collected papers was already under way at MIT Press, and the friends decided to publish a companion volume. It would review and trace the development of Black's ideas. "The authors of the reviews will be contemporaries of yours who can offer firsthand descriptions of the initial impact of your research on academics and finance practice," they wrote.

Black was deeply touched by the planned tribute. On July 19, five days after receiving the letter, he put his appreciation into words. He expressed "stunned surprise" that his friends and colleagues chose to honor his contributions and great pleasure about the manner in which they chose to do it. "I'd rather have this than any other prize out there," he wrote. He accepted his fate stoically, putting the finishing touches on his last papers and preparing his own funeral. Toward the end, he no longer accepted food and declined intravenous feeding. On August 30, 1995, at age fifty-seven, he passed away.

———————

For years it had been rumored that the discoverers of the options pricing formula would someday win the highest honor to which a scientist can aspire: the Nobel Prize. After all, solving a riddle that had been around for close to a century was a breakthrough on par with any discovery in physics, chemistry, or medicine. The three men knew that they were in line. The "other prize out there" that Black mentioned in his thank-you letter was of course the Nobel Prize or, more exactly, the Sveriges Riksbank Prize in Economic Sciences in Memory of Alfred Nobel.

Since Nobels are not awarded posthumously, only Scholes and Merton remained in the running after Black's death. They were both working at Salomon Brothers and teaching, Scholes at Stanford, Merton at Harvard. In 1993 Merton left Salomon to join colleagues in founding a firm they called Long-Term Capital Management (LTCM). Scholes joined them a year later. For a while they both kept their academic appointments, but Scholes chose to become professor emeritus in 1996 in order to devote himself full-time to LTCM. On October 14, 1997, the phones finally rang.

Later during the day a press release announced that "the Royal Swedish Academy of Sciences has decided to award the Bank of Sweden Prize in Economic Sciences in Memory of Alfred Nobel, 1997, to Professor Robert

C. Merton, Harvard University, Cambridge, USA and Professor Myron S. Scholes, Stanford University, Stanford, USA for a new method to determine the value of derivatives." The laudation did not ignore Fischer Black's crucial role in developing the theory. It read: "Robert C. Merton and Myron S. Scholes have, in collaboration with the late Fischer Black, developed a pioneering formula for the valuation of stock options. Their methodology has paved the way for economic valuations in many areas. It has also generated new types of financial instruments and facilitated more efficient risk management in society."

The December 10 ceremony and the following banquet were grand affairs. In the presentation speech, the Swedish economist Bertil Näslund from the Royal Swedish Academy of Sciences pointed out that derivative financial instruments, like stock options, serve a highly useful purpose in society by redistributing risks to those who are willing and able to take them. He then singled out the problem that had been an obstacle in the search for the options pricing formula for so long: what risk premium should be used in the evaluation. "The answer given by the Prize-Winners was: no risk premium at all! This answer was so unexpected and surprising that they had considerable difficulties in getting their first articles accepted for publication." Underscoring the importance of options apart from markets like the CBOE, he suggested a somewhat unexpected example. "The possibility of switching from one type of energy to another is an option," he pointed out, "and the economic value of flexibility can now be determined."

At the black-tie banquet in the evening, Merton, addressing the king, the royal family, the cowinners, and the guests, closed a circle by thanking his intellectual predecessors. "Like all scientists, we too stand on the shoulders of giants . . . ranging from the pioneering French mathematician, Louis Bachelier, to the all-knowing U.S. economist, Paul Samuelson."

———

Merton and Scholes had reason to feel satisfied. They had just climbed the highest rung of the science ladder and were enjoying a lavish dinner with the king of Sweden. Financially, they weren't doing too badly either, and not just because the Nobel came with a seven-figure award. As principals and cofounders of LTCM, a firm that would undertake arbitrage on a global basis, they were making money. LTCM's goal was to marry the best of finance theory with the best of finance practice. It was the quintessential hedge fund.

The firm had begun active business in February 1994. By the time the Nobel Prizes were awarded to the two principals and cofounders in 1997, the firm had 180 employees, three offices—in London, Tokyo, and Greenwich, Connecticut—and about $4 billion under management. "The distinctive LTCM experience, from the beginning to the present, characterizes the theme of the productive interaction of finance theory and finance practice," Merton proudly proclaimed. Scholes seconded. "By applying financial technology to practice, I have achieved a better understanding of the evolution of financial institutions and markets, and the forces shaping this evolution on a global basis." The journal *Institutional Investor* characterized the collection of people at LTCM as "the best finance faculty in the world."

Little did the two Nobel laureates know what was brewing.

the great global colony collapse, aka great recession

SEVENTEEN

... The Harder
They Fall

T HE ÉMINENCE GRISE OF LONG-TERM CAPITAL MANAGE-
ment (LTCM) was John Meriwether, a legendary bond trader
who had built a hugely successful fiefdom inside Salomon
Brothers in the 1980s. After leaving Salomon Brothers in 1991 under a
cloud, he and some key colleagues created LTCM to undertake fixed-
income arbitrage on a global basis. Happy to put their financial theories
to the test (and make a buck on the side), Robert Merton and Myron Scholes joined Meriwether's endeavor. The firm put together financial,
telecommunication, and computer technologies, hired strategists and operations people to run them, executed complex contractual agreements
with investors and counterparties, and raised over a billion dollars from
investors. "It was deliciously intense and exciting to have been a part of
creating LTCM," Merton would recall in an autobiography. "For making

it possible, I will never be able to adequately express my indebtedness to my extraordinarily talented LTCM colleagues."   .

In retrospect, Merton's gushing description sounds like a bad joke, and his "indebtedness" would extend way beyond his colleagues. Together with their "extraordinarily talented colleagues," Merton and Scholes managed to create the biggest flop Wall Street had ever known. It nearly brought down the American banking system.

The son of an accountant, John Meriwether grew up in an Irish Catholic neighborhood on the south side of Chicago. After earning his bachelor's degree at Northwestern University on a scholarship for golf caddies, he taught high school math for a year and then went on to obtain an MBA at the University of Chicago. In 1974, the twenty-seven-year-old Meriwether joined Salomon Brothers. There he created the most successful team Salomon ever had—the bond arbitrage group.

Staffed with smart recruits, this tightly knit, secretive group soon became the most profitable unit within Salomon Brothers. In recognition of the group's success, Meriwether was named vice chairman of the firm in 1988. The honeymoon did not last long. In 1991 one of his employees ran afoul of bond auction rules instituted by the U.S. Treasury, and Meriwether was forced to resign. Even though he himself was not directly involved in the fraud, he had concealed the misdeed out of loyalty to his associate. As a sanction, the Securities Exchange Commission imposed a $50,000 fine and a three-month suspension.[1] When the time was up, Salomon's new chairman—the previous chairman, John Gutfreund, had also been forced to resign in the wake of the scandal—offered him the opportunity to rejoin the firm, albeit with reduced authority. Meriwether, who had considered himself innocent of any wrongdoing, could not accept being disgraced and decided on a change of setting. He founded LTCM.

He was joined by ten associates, nine of them colleagues from Salomon Brothers. They were Scholes and Merton, the two Nobelists; Eric Rosenfeld, Robert Merton's former Ph.D. student and later a Harvard professor who had initially enticed Merton to join Salomon; Victor Haghani, a graduate from the London School of Economics who had started his career in the research department at Salomon Brothers before joining Meriwether's arbitrage group; William Krasker and Gregory Hawkins, also Ph.D.s from MIT; Larry Hilibrand, once the highest-paid trader at Salomon, who took home $23 million in one year; James McEntee, Meriwether's golf buddy; and Richard Leahy. The only Salomon "outsider" among the founding partners was David Mullins, also an MIT Ph.D. and professor at Harvard who had served as vice chairman of the Federal Reserve System. Seven of the eleven founding partners had doctorates, all of them, except Scholes, from MIT. Five of them were or had been finance or economics professors, four of them at Harvard. (The odd man out was Scholes again, who was teaching at Stanford.)

These eleven founding partners decided to create a "hedge fund." The term requires some explanation, especially in terms of how it differs from common mutual funds. The latter present investment opportunities for those who want to set some money aside, hoping for moderate growth. Often these small-scale investments are meant to guarantee sufficient funds for college or retirement. Since most investors cannot be expected to understand the intricacies of financial markets, mutual funds are heavily regulated. Hence investment opportunities are quite limited. They mainly invest in stocks, straight and simple, and the fund managers' supposed expertise consists only in picking the right ones. In some years, one mutual fund may be better at stock picking than the others, in other years the ranking may be reversed. In the long run, most mutual funds perform no better and no worse than the stock market overall.

Hedge funds, on the other hand, are unregistered partnerships, restricted to at most ninety-nine investors. Two-thirds of these limited partners must have a net worth of more than a million dollars. There are only a few restrictions on trading strategies. Hedge funds buy and sell stock, commodities, real estate, options, and other derivatives. They are allowed to sell short. They can borrow as much as lenders are willing to give them in order to leverage their investments. These private investors' clubs are run by general partners who usually charge a management fee of 1 percent of the amount under management, plus 20 percent of any profits.[2]

Since hedge funds are unregistered, their investors do not enjoy the protection of the Securities and Exchange Commission. On the other hand, the profits afford the investors some tax advantages. Hedge funds are not allowed to advertise and are very secretive about their trading strategies. The invested funds are also very illiquid. Partners are usually permitted to withdraw money only once a year, and only after giving many months' notice.

The designation "hedge fund" may erroneously evoke the impression of a safe investment, like a hedged bet that carries no risk. Even though there is some justification for this perception, as I will presently explain, the term is actually a misnomer. By investing borrowed money, hedge funds are exposed to significant risk. While mutual funds are exposed 100 percent to market risk, hedge funds may be exposed 200 or 300 percent, or even more, to the vagaries of the market, depending on how much money is borrowed. Hedge funds are playgrounds for the rich who want to become even richer but would not be impoverished if their investments turn sour.

With such a stellar lineup of partners, LTCM was sure to attract investors. The founding partners put up $100 million and within a short time eighty limited partners, banks as well as wealthy individuals, had paid in an additional $1.25 billion. Among them were the Hong Kong Land & Development Authority, the Bank of Taiwan, the Bank of China, Dresdner Bank of Germany, the Bank of Bangkok, UBS from Switzerland, Italy's central bank as well as the heads of Merrill Lynch, Paine Webber, Bear Stearns, and many others. Each of these limited partners had to invest at least $10 million, which could not be withdrawn for at least three years. Even then the investors could not pull out more than 12 percent of the capital at a time. With thirty employees, $20 million in computer equipment and leading-edge mathematical trading models, LTCM opened for business. Given their larger-than-life opinion of themselves, LTCM's general partners charged a 2 percent management fee and 25 percent of all profits that would accrue.

LTCM's investment strategy was as beautiful as it was simple. Basically, it went by the Black, Scholes, and Merton theory of options pricing. Recall that in order to derive the correct price of an option, the three discoverers of the pricing formula had created a virtual portfolio, composed of stock and options on this same stock. By holding one long and a suitable number of the others short, the portfolio could be rendered completely riskless. If the value of one part of the portfolio went up, the value of the other went down, and vice versa. The hedge fund's task was to combine pairs of stocks and options into riskless bundles.

But LTCM wanted to make profits, not just create riskless portfolios. Armed with the options pricing formula, the partners could compute the correct value of all options. The secret was to find the ones that were mispriced. What they needed to do was to check the actual prices at which options were being traded on the Chicago Board Options Exchange, or at any other bourse, and identify any that did not conform to the theoretical values predicted by the formula. As soon as a mispriced option was found, they did one of two things: if the option's price was lower than the formula said it should be, they quickly bought it and sold the appropriate number of shares short. If the option was priced too high, they sold these options short and simultaneously bought an appropriate number of shares. Not only was the resulting bundle riskless, but since the option was mispriced its price would sooner or later have to adapt to its real value. Thus LTCM was able to make a profit without taking on any risk.

For a while, the strategy worked like a charm. Since many traders and investors were not yet knowledgeable about the options pricing formula and how to apply it, they often determined the prices by rule of thumb or gut feeling. Consequently many options were mispriced. Whenever LTCM's computers identified some whose prices did not conform to the correct value, as given by the pricing formula, traders pounced. Hence, for a while, LTCM could profit from other traders' ignorance.

But while LTCM's strategy followed financial theory to the dot in one realm, it violated another. As pointed out in Chapter 13, the Chicago School's theory of efficient markets claims that the prices at which securities are traded reflect all publicly available information. By now, two decades after the papers by Black and Scholes and by Merton had been published, the formula should certainly have been common knowledge. Recall that Texas Instruments had rejected Merton's demand for royalties by arguing that the formula was in the public domain. Hence the more savvy the traders and the general public became about the theory, the more the prices of options would reflect their true value, and the more difficult it would become to find options that were sufficiently mispriced to cover transaction costs and still leave a profit.

The guys at LTCM would not have been so "extraordinarily talented" if they did not have other tricks up their sleeves. One trick was to exploit differences in the yield spread of U.S. Treasury securities. To illustrate: a three-year Treasury note and a nine-year Treasury bond issued six years earlier carried the same, albeit infinitesimally small risk that the U.S. Treasury would default on its loan. The date of their maturity was also identical. Hence the two securities should have provided exactly the same yield to the investor. However, this was true only theoretically. In practice, T-notes that had just been issued—that are "on the run" in trader lingo—were more actively traded than their older ("off the run") cousins. Should the need arise, investors could offload them more easily, which meant that they usually sold at slightly higher prices. (The resulting higher yield of "off the run" T-bonds compensated the investor for their relative illiquidity.) Typically the yields converged within six months of the issuance of the younger bond. At that point in time, the formerly younger bond would be considered off the run, because by then the Treasury would have issued an even newer bond. Thus LTCM could exploit the minute yield difference by buying the higher-yield off-the-run security, selling the more expensive on-the-run security short, and waiting for the yields to converge. The deal would be totally riskless and provide an assured arbitrage profit.

Other such deals abounded. For example, in 1999, Italy—together with ten other countries—was set to adopt the euro as its currency. Henceforth, loans would be denominated in euros and the high Italian interest rate would have to adapt itself to the interest rates in force in the other European countries. One of the partners soon spotted this misalignment as an opportunity for arbitrage profits.

There was a problem with such trades, however: potential profits were minute, often a few hundredths of 1 percent. The profit on a trade of a

------

* See footnote on page xiv.

$1,000 in T-bonds could yield maybe 20 cents. To make any kind of profit, trades would have to amount to billions of dollars. A purchase of $1 billion in T-bonds would yield a more respectable profit of $200,000. This is why the partners compared their investment strategies to a giant vacuum cleaner "sweeping up pennies."

But not to worry, LTCM had a way around that problem. It simply borrowed lots of money in order to make huge deals. This was not at all unusual; many hedge funds used their assets as collateral in order to leverage their profit potential. What was unusual was the staggering use LTCM made of this possibility. With $4–5 billion in assets, LTCM was able to borrow twenty to thirty times as much. Given the illustrious names of the partners, banks were only too eager to grant loans to LTCM.

Since LTCM's bread and butter consisted in exploiting market imperfections, the firm had to try to maintain the existence of mispricings on the stock exchanges. But the theory of efficient markets—to which the LTCM partners subscribed, of course—claims that security prices reflect all publicly available information. As a consequence, mispricings would immediately disappear. Therefore, in order to allow the goose to continue laying the golden eggs, any information the partners had about mispriced options had to be kept from the public. The direction of their trades, for example, would have permitted surreptitious observers to glean information about the riskless bundles LTCM was trying to create. For that reason, everything LTCM did was steeped in utmost secrecy. Since at least two trades were necessary to create riskless bundles—a purchase of shares, say, and a short sale of options—LTCM traders camouflaged their tracks by placing the two orders through two different channels. Thus nobody outside the firm was ever aware of what LTCM was up to.

The firm's performance was nothing if not spectacular. Right from the start, the mathematical trading models—not only the Black-Scholes equation but also others that had so far been tested only in virtual trades—produced amazing results. In the first ten months of its operation, from February to December 1994, LTCM generated nearly 20 percent profits for its investors, even after deducting its enormous fees. Note that this occurred in a year in which stocks grew by a measly 1.3 percent. The next year was even better, with a return of 43 percent after fees. And in 1996 profits were close to 41 percent. In 1997 returns were down somewhat, but the firm still generated 17 percent after deduction of all fees. The partners could rejoice. Within less than four years, those who had been with LTCM from the start had nearly tripled their investments.

And then the tide started to turn. Another tenet of financial theory is that the volume of any investor's trades is so small that it won't affect the market. Even without making it explicit, the assumption is that whenever a price is posted, traders can buy or sell as many securities at that price as they want. This is actually true for you and for me and for the proverbial man on the street. But LTCM's trades were huge and did affect prices, sometimes significantly so. Furthermore, other traders started to catch on and sophisticated mathematical trading models were no longer the exclusive domain of LTCM's Ph.D.s. Thus markets were becoming more efficient and possibilities for arbitrage, rare. There was just too much capital chasing too few profit opportunities. At the end of 1995 LTCM closed the fund to any new outside investments, and toward the end of 1997 the firm actually returned $2.7 billion, more than a third of the assets, as dividends to the investors.

The fact that profit opportunities were drying up constituted a problem, but it was not a disaster. It was another six months before the sewage really hit the fan. Investors who had their money returned and hopefuls who had been refused entry into the exclusive club would turn out to be the lucky ones.

At the height of its success in 1998, LTCM had $7.5 billion under its management and employed a staff of nearly two hundred associates and back office collaborators. A decade later, in a lecture at MIT in February 2009, Eric Rosenfeld, one of the partners who had been with the fund from its inception, described the events that brought down the behemoth in just five short weeks. In a dispassionate voice and with much hindsight, the former assistant professor of economics at Harvard with a Ph.D. from MIT described the fund's final trials and tribulations.

Problems started in August. The outlook on the Russian economy was bleak and Russian bonds were selling at a discount. Consequently the yields were unusually high, while U.S. Treasury bonds were trading at regular levels. The partners at LTCM were convinced that the spread had to narrow. Hence they took a position in Russian bonds and offset it with a short position in U.S. Treasury bonds. They did not care whether interest rates in general went up and all bond prices down or whether interest rates went down and bond prices up. After all, it is the essence of arbitrage to offset risk by hedging the position. Thus LTCM was hedged against uniform price moves in the bond market. What the partners bet on was that the spread would narrow.

But this time things did not work out as the models predicted. General investors, wary of worsening conditions in Russia, took a flight to safety. They invested in safe T-bills. And the higher the demand for T-bills, the higher their price and the lower their yield. For the Russian bonds, the situation was exactly reversed. Nobody wanted them, their price dropped further, and consequently the yield soared. Thus the spread between the Russian and the American yield widened. It was the opposite of what LTCM had banked on.

On August 17, Russia devalued its currency and defaulted on ruble-denominated bonds that Goldman Sachs had floated just six months earlier. It was the very first coupon payment, and Russia missed it. A decade later, Rosenfeld was still incredulous. "All you needed was a printing press," he said. "Print out rubles and you make the payment. Were there no printing presses in Russia?" LTCM lost on the Russian bonds versus U.S. T-bills deal. But that was not the worst of it. The Russian default had only been the trigger. What really mattered was that as a consequence of the debacle, yield spreads all over the world widened. With investors fleeing to safer havens, most financial instruments moved in parallel, thus nullifying the risk-reducing effect of diversification.

Now LTCM was really beginning to hurt. Nearly all its trading positions were losing value. On Friday, August 21, the fund lost a whopping $550 million. This was five times as much as the partners had imagined in their worst nightmares. The mathematical models that had been built by the best and the brightest from MIT and Harvard predicted that in a worst-case scenario, LTCM would never lose more than $105 million on a single day.[3]

And the losses continued. In August, the fund's net asset value (assets minus liabilities) experienced a loss of 44 percent and was down about 50 percent for the year. Negative rumors began to circulate and the possibility of bankruptcy arose. Consequently funds that had trading positions similar to LTCM's began to liquidate their holdings in order to minimize their losses in case the firm was forced into bankruptcy. Other market participants established positions that would benefit from the firm's forced liquidation. "We were like a big ship in a small harbor, trying to turn. Everybody was trying to get out of the way."

As investors tried to liquidate their holdings in weak markets, prices were driven down. And since they were doing that simultaneously, everything became correlated. Even though different securities were not economically correlated, prices moved in parallel because of the general panic. The age-old strategy of diversification, based on the statistical fact that spreading holdings over many unrelated securities minimizes risk, failed miserably. Moreover, as investors pulled back simultaneously, markets dried up. All this prevented LTCM from rapidly scaling down its positions. The fund was stuck with a highly toxic portfolio.

Monday, September 21, was another memorable day. LTCM lost $551 million, an amount eerily similar to the loss on its previous worst day. That was too much even for the Fed, which had until then remained on the sidelines, watching the free markets take their toll. On Wednesday, September 23, William McDonough, the president of the Federal Reserve Bank of New York, called a meeting of some fourteen leading investment houses and commercial banks. He wanted to arrange a bailout.

McDonough was afraid that if LTCM went bankrupt, pandemonium would result. Remember that LTCM always hedged its trades by creating riskless bundles. When shares were bought, options were sold short, when nine-year T-bills were shorted, three-year T-bills were bought, and so on. In order to reduce risk to a minimum, LTCM had a countertrade for every trade on its books. So even though the value of the portfolio had gone down drastically, when considered as a whole, there was little risk in the thousands of positions that LTCM held. But in order to keep their investment strategies secret, LTCM had done the trades with different banks. The Fed's president feared that this could now come to haunt the whole banking system. After all, the trading partners saw only their own exposure to risk. They were ignorant of who had made the countertrade, or even what it was. If and when disaster hit LTCM, they would want to salvage their position by buying or selling securities at whatever prices they could get. The banks would be the losers, not LTCM, and that is what worried the Fed. On the other hand, if the major trading partners got together and analyzed the situation from a global viewpoint, they would understand that altogether there was much less risk in LTCM's positions than each one was holding on its own. Once they realized that, McDonough hoped, they would be willing to bail out the ailing LTCM.

The meeting started at 10:00 AM. Wall Street's most important bankers attended. Associates and lawyers had been working through the night. But at 10:30 McDonough suspended the meeting. A message had come through by satellite phone from Alaska. It was a last-minute offer from Warren Buffett, who had recognized the potential and was willing to invest. He was vacationing in Alaska with Bill Gates when he placed the call to Meriwether. A fax detailing the proposal arrived half an hour later: $250 million in exchange for the assets of LTCM. Buffett, together with Goldman Sachs and the American International Group (AIG), would then put up the $4 billion necessary to save LTCM. Of course the general partners who had driven the fund into the ground would be fired. The offer was good for one hour!

Meriwether and his partners went into a huddle to consider the offer. It was a far cry from the $4.8 billion that the portfolio had been worth just nine months ago, but "why quibble over a few dollars," as Rosenfeld put it; $250 million was better than zero. Unfortunately, their lawyers came to the conclusion that the deal could not be done. LTCM had 15,000 outstanding trades. Moving them to a different counterparty required the original trading partners' approval. And all they had was an hour. They tried to reach Buffett to explain the problem and make a counterproposal. But he could no longer be reached in the wilderness. So the deal fell through.

At 1:00 PM the bankers, under the Fed's guidance, resumed their discussions. McDonough coaxed and cajoled. By late afternoon, after many angry exchanges and much see-sawing, an agreement was reached. A consortium of fourteen institutions, the mightiest banking establishments on Wall Street as well as in Switzerland, Germany, and Britain, agreed to invest $3.625 billion in the fund. The original investors kept their stake, now worth $400 million. Though many of those present in the Fed's conference room wanted to fire the LTCM partners who had produced the mess, they were persuaded to keep them on. They were needed to liquidate the reanimated fund.

Much criticism followed the Fed's participation in the events. McDonough took pains to emphasize that although the Fed had brokered the deal, no public money had been employed to prop up the failing firm. The $3.625 billion came entirely from banks. But the Fed's involvement had other implications. By creating a mind-set among executives of important financial institutions that their businesses were "too big to fail," the orchestrated, even if privately financed, rescue of LTCM may have planted the seeds for the next crisis. In the events leading up to the financial meltdown of 2007, memories of the bailout may have induced Lehman Brothers and Bear Stearns to take on more risk than they should have, in the mistaken belief that the Fed would always be there to help out.

The members of the consortium did not fare too badly. Helped by recovering financial markets, LTCM's portfolio increased in value. The portfolio was liquidated during the following months, and overall the members of the fourteen institutions made about 10 percent on their investment.

The fate of the partners can be viewed in two strikingly different ways. One way of looking at their fortune is that $1 at the beginning of 1998 fell to 8 cents at the height of the crisis and was liquidated for about 10 cents in the middle of 1999. This is really bad, you might say.[4] However, a fairer way of evaluating LTCM's performance is to consider the entire period of its operation. And here the picture looks quite different. Taking into account the $2.7 billion dividend at the end of 1997, investors who had been with the firm from the outset had made a very respectable 19 percent per year, on average. Moreover, of the hundred limited partners only a dozen lost money and only half of these lost more than $2 million. Of course, some were badly hit. Leader of the pack was the Union Bank of Switzerland, which lost close to a billion Swiss francs. ∪ ฿ S !

If we believe Rosenfeld, only the general partners really lost money, close to $2 billion, as he plaintively maintained in his MIT lecture. He conveniently forgot to mention what he had claimed just a minute before—that performance should not be measured from the fund's high point to its low point but from its inception to its end. Viewed in this way, the general partners profited healthily from the 2 percent management fee and 25 percent profit sharing during the four fat years.

Minor obstacles such as a multibillion dollar loss were not going to deter the partners for long. In December 1999, after LTCM had been wound down, Meriwether, together with Rosenfeld, Haghani, and Hilibrand, started a new hedge fund, JWM Partners. They used the same offices and the same models as LTCM—and even some of the same clients. This time, however, they instituted a more conservative approach: stricter oversight, more controls over risk, and limitations on its borrowings. At least that is what the principals claimed. During the financial crisis of 2008, ten years after LTCM had gone up in smoke, Meriwether was scrambling again. Between September 2007 and February 2009, the fund's value decreased by 44 percent. In July, just as other hedge funds were starting to recover from the latest financial crisis, JWM Partners announced that it was shutting down its Relative Value Opportunity II fund. Whether Meriwether ever gets Opportunity III is doubtful. But at least JWM Partners had lasted twice as long as LTCM.

Myron Scholes did not stay outside the arena for long. With one of his students from Stanford, Chi-fu Huang, who became professor at MIT and took one of his students along, Scholes founded Platinum Grove Asset Management LP in 1999. Platinum Grove generated profits of nearly 10 percent for its clients, after fees, until 2007, with low volatility and no correlation with the bond market or the stock market. But in November 2008 (what did you expect?) the fund was about to blow up. After being down 38 percent for the year, having lost 29 percent in the first half of October alone, the fund stopped investor withdrawal. "The suspension is necessary given current market conditions," the fund explained in an e-mail statement. "Platinum Grove will use this period to consult with its investors and counterparties, determine their future intentions and manage the assets of the fund accordingly." LTCM all over again? Not exactly. At least the company still had a website in 2011.

Robert Merton meanwhile cofounded Integrated Finance, a financial advisory firm, in December 2002. Unfortunately, no clients were interested in getting advice from the former LTCM rocket scientist. In 2007 the firm was merged into Trinsum Group, with Merton as chief science officer. On January 30, 2009, Trinsum filed for Chapter 11. The company specified liabilities of $15.8 million against assets of $1.2 million. Of the latter, $1.1 million was listed as the value of a patented software . . . a meek postscript to the Nobel Prize–worthy research that had led to the solution of the century-old problem of options pricing.[5]

———

A few lessons may be learned from LTCM's failure. Trades that require leverage because they yield only fractions of a percent were now more accurately likened to "a giant vacuum cleaner sweeping up pennies . . . in front of a steamroller." Never forget that the statistical distributions of economic data have what is called fat tails. Events that are so rare that theoretically they should occur, say, once in a millennium, do happen time and time again. And when disaster does hit, diversification is worthless because everything correlates with everything else and markets dry up. The main lesson of the LTCM disaster, however, is that secrecy, combined with high leverage and dangerous doses of greed and arrogance, can add up to a highly toxic mixture to which even the smartest doctors, Ph.D.s that is, have no antidote.

EIGHTEEN

The Long Tail

C APITAL MARKETS AND THE MARKETS FOR DERIVATIVES
form the foundations of a modern economy. By matching entrepreneurs who require money to operate and expand their
businesses and investors willing to put their savings at the businesspeo-
ple's disposal, capital markets allow an economy to develop and flourish.
The so-called derivatives markets, where financial instruments, first and
foremost options, are traded, bring together investors with different tastes
for risk. Risk-averse investors use options to transfer some or all of their
investment risk to trading partners who have the stomach to assume it.
The combination of the two types of markets allows investors to build
portfolios that reflect their preferred mixture of investment and riskiness.
Of course, those who purchase risk need to be rewarded for their willingness to carry it. For centuries the appropriate compensation was determined by supply and demand. The counterparties somehow agreed on a
price. But the whiff of improvisation that surrounded such trades prevented the options markets from becoming fully developed. No records
exist about the number of times that potential trading partners walked

away from deals because they could not agree on a price. Hence there was a pressing need to assess an option's true value.

When they discovered the options pricing formula in 1973, Black, Scholes, and Merton provided an intellectual milestone. But they did more than that. Their discovery caused markets for derivatives to become more efficient and significantly advanced their ubiquity. Nowadays, options are traded at exchanges all over the world, from Mumbai, to Shanghai, to Zürich, to Chicago.

Of course, there are pitfalls. I will mention three. A model may be elegant, a formula beautiful, but if the input is nonsense, the output can be no better. "Garbage in, garbage out," as the saying goes. One variable in particular, the stock's volatility, is a crucial input for the Black-Scholes equation. In general, historical volatility of the share prices is used in lieu of future volatility. But volatility is volatile. Hence past data may not be the best predictor of the future.

The need for the continuous adjustment of the portfolio—the ratio of shares held short for each option—is another pitfall. In order to keep the portfolio riskless, the model requires that the hedge ratio be adjusted as time progresses and, more importantly, whenever the price of the underlying share changes. With stock prices performing Brownian motion, the portfolio would have to be continually rebalanced. But even if it were possible to trade shares minute by minute, it would entail enormous transaction costs. After all, in spite of the model's assumption that there are no transaction costs, brokers do want to be paid.

The most frustrating pitfall is probably the normal distribution, even after Osborne's and Samuelson's modification of Bachelier's original model. One of the hallmarks of the normal distribution is that the vast majority of the events it describes fall within a relatively short distance around the mean. If you picture the bell-shaped curve, then 99.7 percent of all events are contained within the body of the bell, and only 0.3 percent occur in the tails on the right and left side. (For the technically inclined: I here define the tail as being anything beyond about three standard deviations from the mean.)

Let's say that the average daily movement of a stock index is zero and the standard deviation is half a percent. According to the normal distribution, 68 percent of all movements would be less than half a percent in either direction, 95 percent less than 1 percent, and only 0.3 percent more than 1.5 percent up or down. Thus, on average, one such extreme movement should occur about once a year. Most of the time—for small- and medium-size movements—the bell-shaped distribution describes movements on the stock exchange quite well.

But why do we care about such infrequent events? Aren't we overstating their importance? The answer is no. The really interesting things in the stock market, like booms and busts, are not described by the figure's central part, but by its tails. And this is where the normal distribution leads us astray. Supposedly rare events occur more frequently than the bell-shaped curve suggests. Unexpected and unpredictable events, like the sighting of a black swan, several successive droughts, or Russia's default in August 1998, do happen. For example, while there was only a single day between 2004 and 2006 on which the Dow Jones Industrial Index moved by more than 2 percent, only a year later, in 2007, it happened fourteen times! Apparently, extreme movements occur more frequently than the normal distribution predicts. The conclusion is that the normal distribution is not a good description of stock price movements. More weight must be given to extreme events; the distribution of stock price movements obviously has fatter tails than the normal distribution.

So there are pitfalls and they can and do lead to spectacular failures like the LTCM crash, the crisis of 2007, and the demise of Lehman Brothers. But to fault Black, Scholes, and Merton and their equations for such aberrations would be utterly wrong. This would be like accusing Isaac Newton and the laws of motion for fatal traffic accidents.

Risk was a keyword throughout this book. The willingness to assume more risk in exchange for higher profits, or to accept lower profits in exchange for less risk, is what drives investments on the stock exchange. This has been known for centuries. It took the skills of Black, Scholes, and Merton, however, to put these insights into neat models. With advanced mathematical techniques these modern-day quants managed to dissect and explain a phenomenon that had been observed for decades but never fully understood.

One result of their effort is a thriving derivatives market where investors can buy and sell risk. But caveat emptor! Superficial application of misunderstood models, disregard for their limitations, and simplifying assumptions and boundary conditions may create the false illusion that risk has been eliminated. Unwary investors, even financial institutions, may be led to ruin. Nevertheless, whatever the practical, and sometimes unfortunate, consequences, on a scientific level the development of the options pricing formula is a landmark achievement of the twentieth century.

A Pedestrian's Guide to the Derivation of the Black-Scholes Equation

Here I give a very simplified and nonrigorous description—just a flavor, really—of the derivation of the options formula.

If we write M for the number of shares, S for the share price, N for the number of options, and C for the option price, then the value of the portfolio, V, can be expressed as

(1) $\quad V = M \cdot S + N \cdot C.$

When the prices of the share and the option change, the value of the portfolio also changes (changes are denoted by a prefixed d):

(2) $\quad dV = M \cdot dS + N \cdot dC.$

According to Itō's calculus (see Chapter 11), dC can be developed into a so-called Taylor series, where σ^2 denotes the stock's volatility, and ∂ denotes a partial derivative (which will not be defined here):

(3) $\quad dC = \dfrac{\partial C}{\partial S} dS + \dfrac{\partial C}{\partial t} dt + \dfrac{1}{2} \dfrac{\partial^2 C}{\partial S^2} \sigma^2 S^2 dt.$

Replacing dC in equation (2) with the right-hand side of equation (3), the change in the value of the portfolio becomes,

(4) $\quad dV = M \cdot dS + N \cdot \left[\dfrac{\partial C}{\partial S} dS + \dfrac{\partial C}{\partial t} dt + \dfrac{1}{2} \dfrac{\partial^2 C}{\partial S^2} \sigma^2 S^2 dt \right].$

Now let's adjust the ratio of "shares to options" in such a manner that the portfolio is hedged—that the value of the portfolio remains unchanged after a price change. This ratio must be the inverse of how the price of an option changes in response to a change in the price of the share:

(5) $\dfrac{M}{N} = -\dfrac{\partial C}{\partial S}$

Hence if the price of the option rises by, say, 50 cents in response to a \$1 increase in the share price, $\partial C/\partial S = 0.5$, then two options must be held short (hence the minus sign) in order to hedge the portfolio.

Let's look at the portfolio that contains just one share, i.e., we set $M = 1$. By equation (5), the number of options needed to hedge the portfolio is,

(6) $\quad N = -\dfrac{1}{\partial C/\partial S}.$

Replacing N in equation (4) by this ratio, the change in the value of the hedged portfolio becomes,

(7) $\quad dV = -\left(\dfrac{1}{\partial C/\partial S}\right)\left[\dfrac{\partial C}{\partial t} + \dfrac{1}{2}\dfrac{\partial^2 C}{\partial S^2}\sigma^2 S^2\right]dt.$

(Note that $M \cdot dS$ and $N \cdot \partial c/\partial s \cdot dS$ cancel each other because $M = 1$.) Since this portfolio is hedged—riskless—its return in percent during a time period, dV/V, must be equal to the risk-free interest rate, denoted by r, multiplied by the elapsed time period, dt:

(8) $\quad dV/V = r \cdot dt.$

Rearranging terms and substituting equation (1) for V and equation (6) for N (recalling that M=1), we obtain:

(9) $\quad dV = r \cdot dt \cdot V = r \cdot dt\left(S - \dfrac{C}{\partial C/\partial S}\right).$

Setting equations (7) and (9) equal and manipulating the terms, we obtain the differential equation that stumped Black and Scholes:

(10) $\quad \dfrac{\partial C}{\partial t} = rC - rS\dfrac{\partial C}{\partial S} + \dfrac{1}{2}\dfrac{\partial^2 C}{\partial S^2}\sigma^2 S^2.$

Of course, there is an additional condition, called a boundary condition, which says that on the exercise date the price of the option, C, must be equal to the larger of zero or the difference between the stock price and the exercise price. Equation (10) contains the terms $\partial C/\partial t$, $\partial C/\partial S$, and $\partial^2 C/\partial S^2$. An equation that contains such terms is a differential equation. And differential equations are notoriously difficult to solve.

With a very complicated change of variables, the differential equation (10)—which describes how the option price changes with time—can be changed into an equation that is well-known in physics: the heat transfer equation. It describes how heat propagates through a solid body, say a metal rod. It also has a boundary condition because at the edge of the rod the heat has nowhere to propagate to. Let's say that the heat at position x at time t is denoted by $U(x,t)$. Then the heat transfer equation says,

$$(11) \quad \frac{\partial U}{\partial t} = \frac{\partial^2 U}{\partial x^2}.$$

The solution to this equation was already proposed by Joseph Fourier in 1822 in his book *Théorie de la chaleur* (Theory of heat). It is,

$$(12) \quad U(x,t) = e^{-\frac{xt}{\gamma^2}}\left[A \cdot \cos\left(\frac{x}{\gamma}\right) + B \cdot \sin\left(\frac{x}{\gamma}\right)\right],$$

where γ stands for the unit of length. Reversing the complicated change of variables, solution (12) can be transformed into the solution of the differential equation (10), as given by Black, Scholes, and Merton and depicted in Chapter 15. (See page 219.)

NOTES

CHAPTER TWO

1. I have consulted various sources about John Law, many of which trace their origins to Charles Mackay's account "Extraordinary Popular Illusions and the Madness of Crowds."

2. Madame de Tencin was the mother of the famous mathematician, physicist, and philosopher Jean leRond d'Alembert, whom she abandoned as a newborn on the steps of a church.

3. Galignani, page 221. The restriction on women remained in place until 1967.

4. By the same token, a short purchase implied the laudable expectation on the part of the speculator that the security's price would rise. This observation was apparently lost on the powers that be.

5. Everybody believed that the share prices would rise and differed only over how much. Say the security stands at 100 francs. Monsieur Dupont believes the price will rise to 130 by the next settlement date, and Monsieur Bouvier believes it will rise to 150. They enter a forward contract: Dupont agrees to sell and Bouvier agrees to buy a share of the security at a price of 140. If the price does rise to 150, Dupont must buy the share on the open market at 150 and deliver it to Bouvier for 140. Bouvier makes a gain of 10, Dupont loses 10. If the share price turns out to be 130, Dupont buys the share on the open market for 130 and sells it to Bouver at 140. Dupont gains 10, Bouvier loses 10. Actually, the security need not even change hands. Bouvier and Dupont may simply exchange the 10 francs on the settlement date, without anybody actually buying and delivering the security.

6. Zola was no anti-Semite, however. His newspaper article "J'Accuse" (I Accuse) in 1898 was a turning point in the notorious Dreyfus affair, in which the Jewish army officer Alfred Dreyfus was wrongly convicted of treason. Zola, who accused the highest levels of the French army of anti-Semitism, had to flee to England after publishing the article.

CHAPTER THREE

1. Part of this biography of Regnault follows Jovanovic 2004.

2. Regnault was thinking of Gauss's normal distribution.

3. $d = gt^2/2$, where d is the distance traveled by a falling body, g is the gravitational constant, and t is time.

4. $L = gT^2/4\pi^2$, where L is the length of the pendulum and T is the period of the oscillation.

5. I touched on futures transactions in the previous chapter.

6. About 450 securities were traded on the Bourse de Paris, one-third of them bonds, two-thirds stocks.

7. Which prompted my former finance professor Marshall Sarnat to deadpan, "but if you do keep them in one basket, make sure you keep an eye on the basket."

8. The formula Regnault had in mind is what we today would call the standard error.

9. The price falls to the so-called ex dividend price.

CHAPTER FOUR

1. One of the villains in Emile Zola's novel *L'Argent* (Money) is Jantrou, the editor in chief of the journal *L'Espérance*.

2. One exception to the rule is that commodity prices must include transportation costs to the various locations.

3. Lefèvre makes a halfhearted attempt to extend his analogy to explain how the body comes to terms with random fluctuations. His answer: the heart valve prevents blood from running backward, thus ensuring the orderly flow of the vital fluid.

4. The slopes of the rising or falling parts of the lines can differ from 45 degrees if more than one option is purchased. If the investor owns two options the slope of the line is equal to 2—about 63 degrees.

CHAPTER FIVE

1. This qualification is significant because it rules out crashes from Bachelier's analysis, such as the crises of recent years.

2. Alumni of the École Normale Supérieure or the École Polytechnique.

3. Translated by Murad S. Taqqu in "Bachelier and His Times: A Conversation with Bernard Bru" (2001).

4. The question was of some interest at the time because there seemed to be a paucity of 7s in the latter part of the digits. Only in 1946 was it shown that an error in the computation of the 528th digit invalidated the remaining ones. Bachelier's result showed that the digits, be they of π or the result of erroneous calculations, could be considered *consistent* with randomness. He had stated at the outset that the existence of randomness can never be proven.

CHAPTER SIX

1. Brown's passing opened a time slot at the Linnean Society three weeks later, which allowed the presentation of Charles Darwin's and Alfred Russel Wallace's famous papers on evolution, "On the Tendency of Species to Form Varieties and on the Perpetuation of Varieties and Species by Natural Means of Selection," on July 1, 1858, before thirty or so not overly impressed members of the society.

2. I will resist the urge to present the reader with yet another biography of this "unknown patent clerk" in Berne, Switzerland.

3. He also estimated the Avogadro number, the number of molecules in one mole, at about 4×10^{23}. The best known estimate today is a bit over 6×10^{23}.

4. Let's be a bit more precise. With two consecutive steps we have the possibilities $+1/+1$, $+1/-1, -1/+1$ and $-1/-1$, each with a probability of 25 percent. The mean location of the drunken sailor is $0.25(2 + 0 + 0 - 2) = 0$, i.e., back at the lamppost. By squaring the displacements we can compute the mean squared dispacement, $0.25 (4 + 0 + 0 + 4) = 2$. Hence the mean displacement after two steps is $\sqrt{2}$.

5. After a day, the particle would have moved about a quarter of a millimeter, which could be just about visible by the naked eye. Because the average covered distance increases only with the square root of time, progress is very slow. It would take sixteen days on average for a particle to move a full millimeter, and about four and a half years to move a centimeter.

6. Actually his full name was Marian Ritter von Smolan Smoluchowski, indicating that his family belonged to the Habsburg nobility.

7. The exact formula for n tosses: expected gain/loss $= \sqrt{(2n/\pi)}$.

8. Mandelbrot first investigated fractal processes in 1962, in the context of market prices for cotton and wheat. ·

9. The piezoelectric effect is the phenomenon that some materials generate electricity in response to mechanical stress. Langevin made use of it in the development of an ultrasonic submarine detector.

10. As the *New York Times* reported on December 13, 1911, "the letters cited by Mme. Langevin's lawyer include all those which Mme. Langevin removed surreptitiously from her husband's second apartment where he is said to have constantly met Mme. Curie, and it is hinted that they show that relations of great intimacy existed.

11. Langevin's paper spawned a new mathematical field—stochastic differential equations.

12. Surely you looked for the object of the last sentence, after reading that they allowed The to use the labs . . .

13. To be fair, Svedberg had no knowledge of Smoluchowski's paper.

14. Milton Kerker 1976.

15. Research is still ongoing. Making very careful measurements of the position and velocity of a tiny glass bead suspended in air, four physicists in Texas were able

to measure the instantaneous velocity of a particle undergoing Brownian motion (Li et al. 2010).

16. Apparently thinking that Marian Smoluchowski was a woman, he referred to him as Marie.

17. Moving pictures (i.e., cinematography) had been invented just one decade earlier.

18. According to Svedberg's will, these notes were not be made public until fifty years after his death—February 2021. In 1984 his widow agreed, however, to have excerpts put on exhibit at the University of Uppsala at the occasion of the hundredth anniversary of Svedberg's birth.

19. Kauffman 2003.

CHAPTER SEVEN

1. The expression "random walk," which today is the standard designation for this sort of phenomenon and is used interchangeably with "Brownian motion," was probably coined from the words Pearson used to introduce his letter.

2. It is also the principle of expensive noise-canceling headphones, which this author tends to wear while neighbors are performing renovations to their house.

3. Specifically, the variances of the Brownian motion and of the errors.

4. Lauritzen 2002.

5. Twenty years later, Paul Samuelson would suffer the same fate (see Chapter 13). Wiener's father had been the first tenured Jewish professor at Harvard.

6. This is in contrast to the disco. When the frequency of the strobe light is increased, the observations approach the dancer's smooth movements.

7. A young colleague had not been so lucky. René Gateaux, three years younger than Lévy, had chosen to attend École Normale Supérieure. When the war started, he was also obliged to serve in the army. But while former pupils of École Polytechnique, like Lévy, were recruited to the artillery corps, the graduates of École Normale Supérieure had to serve in the far more dangerous infantry. Gateaux was called up in August 1914. Two months into the fighting, on October 3, 1914, at one o'clock in the morning, the twenty-five-year-old mathematician was killed in the village of Rouvroy, along the Belgian-French border. Many students of Normale Sup perished at the front while Polytechniciens survived.

In the months before his call-up, Gateaux had prepared material for his Ph.D. dissertation, leaving the work unfinished. When Lévy returned from active duty, a professor asked him to collect the material and prepare the posthumous publication of Gateaux's papers. Lévy obliged but went one better. He considerably developed Gateaux's ideas, which proved to be of great value also to his own intellectual development.

8. The study, which was published in 1996 in the prestigious scientific journal *Nature*, became famous. Over the next ten years it was cited 170 times. There was only one problem with it. It was utterly and completely wrong. The long flights, crucial

evidence for Lévy distributions, did not exist. What had happened was that the sensors started recording time when they were turned on at the computer. But from that moment on, it would still take hours until the contraption was fastened to the albatross's leg. And the bird often lingered on its nest for hours if not days before it finally took off on a foraging flight. All the time the sensor was bone dry of course; hence the time span was interpreted to be flying time. And when the albatross returned to its nest, the time until the sensor was removed from the albatross's leg was again recorded as dry time and again interpreted as flying time. Removing these two presumptive trips from the data eliminated any statistical evidence of Lévy distributions.

CHAPTER EIGHT

1. See Chapter 8 in Szpiro, *Kepler's Conjecture.*

2. See Szpiro, *Poincaré's Prize.*

3. Davis and Etheridge 2006.

4. I will not dwell on all terms and definitions and introduce them only when necessary.

CHAPTER NINE

1. For example, a terrorist organization or a criminal may own two accounts and use them to simultaneously buy and sell financial derivatives. Through the first account, which contains the dirty money, a forward transaction is effected that is completely in opposition to market expectations. The second account serves as the counterpart for the deal. On the exercise day, the first account will lose and the second will make money. Thus the losses in the dirty money account will have been transformed into legitimate profits in the clean money account, and the dirty money will have been laundered. Meanwhile, the inevitable transaction costs, chalked up as business expenses by the launderers, keep the banks and brokers happy.

2. Neither Meithner nor Flusser survived the war. Flusser was killed in the Buchenwald concentration camp in 1940, and Meithner died in prison after being arrested by the Gestapo in 1942.

3. Only prices that lead to an exercise of the option are considered. Security prices that result in the option not being exercised are ignored—weighted with zero.

4. With this they quote the fourteenth-century king Edward III who—amid the snickers of his courtiers—picked up a garter that a lady-in-waiting had dropped and bound it on his own leg to cover her embarrassment.

CHAPTER TEN

1. This was especially true in Soviet Russia, where probability was regarded with extreme condescension if not downright hostility. Trofym Lysenko, the notorious sham scientist, dismissed geneticists who, "being unable to reveal the regularities in animated nature, have resorted to the theory of probability. Physics and chemistry

have rid themselves of the accidental," he claimed and added the punchline "science is the enemy of the accidental." Another Soviet scientist had this to say about Karl Pearson: "His system only rests on a mathematical foundation and the real world cannot be studied on this basis at all."

2. The expected value is the value that will appear on average or, better yet, the average of the values that appear.

3. For example: the rational number ⅝ = 0.125 has the numbers 0.1249999999 . . . and 0.1250000001 . . . near it.

4. His father, an agriculturist who took no interest in his son, was exiled during the tsarist regime, became a department head in the Ministry of Agriculture after the revolution, and perished in 1919 during the Russian civil war.

5. In 1940 Kolmogorov bravely contradicted Trofim Lysenko's pseudoscientific views of biology. Lysenko was notorious for his misguided and brutal suppression of Gregor Mendel's genetic theories. When Kolmogorov examined a study favored by Lysenko because the experimental breeding results supposedly did not conform exactly to Gregor Mendel's predictions, he concluded that Lysenko's supposed refutation of genetic theory was invalid because of the sample's small size. Such an affront to Joseph Stalin's all-powerful darling was a brave accomplishment indeed. Lysenko, at whose behest numerous geneticists were executed, promptly responded with an article that would be funny if its author had not been taken so seriously by the Soviet powerbrokers. It also demonstrated, once again, the low esteem in which probability was held by many at that time. "We biologists, however, do not want to submit to blind chance, even though this chance is mathematically admissible. We maintain that biological regularities do not resemble mathematical laws," the czar of Soviet science wrote, concluding that "we, biologists, do not take the slightest interest in mathematical calculations which confirm the useless statistical formulae of the Mendelists." Faced with a real threat to his career, Kolmogorov had little choice but to qualify his genetics-related work on probability and statistics.

At an assembly of a research council, many mathematicians confessed their "sins" and Kolmogorov took pains to point out that his remarks were not meant as a proof of modern genetics but only indicated a statistical confirmation. Following this humiliating U-turn, he stayed away from biology for many years.

6. For more details, see Szpiro, *Kepler's Conjecture*.

7. The jury awarded Poincaré the prize, because even without a solution to the three-body problem, the theoretical advance in his submission was indeed prizeworthy. But in their eagerness the jurors had overlooked a detail: Poincaré's work contained an error. When it was discovered by an editor of the journal *Acta Mathematica*, where the paper was to be published, the Frenchman was deeply troubled. The prize had been awarded and the prize essay was already being printed. He could be certain of the glee of his competitors. Fortunately, after a few months of feverish

work he was able to correct the error. The copies of the journal that had already been distributed were collected—Mittag-Leffler used subterfuge and his considerable diplomatic skills so nobody would note anything amiss—and were pulped. A new issue was printed and sent to subscribers all over the world. Mittag-Leffler asked Poincaré to foot the bill for the second printing, and the relieved mathematician gladly complied even though the cost exceeded the prize money he had received. (See Szpiro, *Poincaré's Prize*.)

8. Kolmogorov's lukewarm retraction of his rejection of Lysenko's sham biology was also motivated by his fear of the KGB.

9. See Szpiro, *Kepler's Conjecture*.

10. Axioms 1 and 2 are a bit technical.

11. Strictly speaking, this is not quite true. On most stock exchanges, price jumps are limited to dimes or full dollars. This can lead to strange phenomena. (See, for example, Szpiro 1998.)

12. The PDE that Kolmogorov found is of the so-called parabolic type, which has important applications in physics. One of its earliest uses was to describe the diffusion of heat. (Recall that Regnault had already used heat diffusion in his thesis to describe stock market behavior.) When the mathematical symbols are translated into plain English, the heat equation simply states that temperature rises or falls everywhere at a rate proportional to the average temperature in the neighborhood: the hotter the surroundings, the faster the temperature will rise in the center. Another application of parabolic PDEs is the so-called Ricci flow, which in 2003 led to the proof of Henri Poincaré's famous, decade-old conjecture. For an account of the history of this feat, see Szpiro, *Poincaré's Prize*.

13. Even before Kolmogorov took an interest in the problem, a whole array of people were involved in studying parabolic PDEs. Among them were Erik Holmgren and the Croatian-born William Feller in Sweden, Sergei Bernstein in the Soviet Union, the Italian E. E. Levi, and the Frenchmen Jacques Hadamard and Maurice Gevrey. Richard Courant at the University of Göttingen in Germany made the solution of parabolic PDEs a research subject at the university's famous mathematics department, and Kolmogorov applied for a Rockefeller grant to spend some time there. (By the time Kolmogorov was ready to travel in April 1933, Courant, who was Jewish, had been dismissed by the Nazis. Subsequently Kolmogorov decided to go to Paris but nothing came of that either. He never obtained a visa from the Soviet authorities.)

CHAPTER ELEVEN

1. Stochastic, from the Greek word "to guess," is another word for random.

2. Until then, most mathematicians—including the great Carl Friedrich Gauss— had believed that a continuous function had to be smooth everywhere, except possibly

at a number of isolated points where it makes sudden jumps. But with his function, Weierstrass proved the exact opposite and blew his colleagues' minds. He also anticipated by a century the concept of fractals, which became an important area of investigation in mathematics and other fields of science.

3. There is actually a mathematical proof for that, but I am not going into it.

4. I will have more to say on geometric Brownian motion in Chapter 13.

5. I need to add a caveat on the dollars versus percentages problem that relates to the question in financial theory of whether "net present value" or "internal rate of return" should be the relevant measure for investors. The answer is that dollars and cents are the relevant criterion, not percentages: an investor should prefer a 5 percent profit on $1 million to a 20 percent profit on $100 because $50,000 beats $20. But for the description of financial markets by Brownian motion, the percentage increase or decrease is relevant.

CHAPTER TWELVE

1. Jakob had published a partial solution to the so-called isoperimetric problem a year earlier, but Johann pretended not to know about it.

2. Fortunately, the Nazis showed no interest in the dusty boxes of the academy when they occupied Paris. Otherwise they could have found some French nuclear secrets.

3. A sad reason for the use of *plis cachetés* that had not been foreseen by the academy when it instituted the practice was the banning of publications by authors on racial grounds. When Paul Lévy had to go into hiding under a false name during World War II because of his Jewish identity, he also took recourse to the procedure by sending sealed envelopes to the academy.

4. One scholarship was from the Alliance Israélite Universelle, the other from a foundation that had been set up with a substantial fortune that the Marquise Marie-Louise Arconato-Visconti, a well-known Parisian socialite, had bequeathed to the University of Paris. The latter stipend being more substantial, Döblin immediately returned the sum he had already received from the Jewish organization.

5. One person to whom Döblin did not hesitate to present his work was Paul Lévy, the eminent probabilist. Once when he visited the professor at his home, Lévy's daughter asked whether he was related to the celebrated author of *Berlin, Alexanderplatz*, a book she was required to read at university. Given the common interests, the two families grew socially close. Lévy became a friend of Alfred's and a mentor to the young Wolfgang, his daughter a companion to Erna.

6. The citation for the *médaille militaire* states that Wolfgang Döblin, "a brave and dedicated solder, always ready to volunteer for even the most exposed positions, secured the telephone connection within the battalion day and night by repairing lines under enemy fire."

7. Döblin had participated in the battle with great distinction for six weeks. Posthumously, he was awarded another Croix de guerre with palm leaves. The min-

ister of the army commended "the very courageous and audacious telephone operator who fought with a complete disregard for danger.... On June 16 in [the village of] Beneng he remained alone, armed with an automatic rifle, in order to provide cover for a group of soldiers as they retreated. He himself only drew back after having delayed the arrival of the enemy motorcyclists from the village, thus enabling the group to take cover in a new position."

8. Alfred had worked for a while in the French propaganda ministry during the war, and then spent a not entirely unsuccessful but unsatisfying stint in Hollywood. At the end of the war, he returned to France and was sent to Germany as a member of the occupying forces. Clad in the uniform of a French army colonel he served as literary inspector (i.e., censor), which did not endear him to his former colleagues. He was disappointed by the political atmosphere that took hold in western Germany after the war but also resisted efforts by the literary establishment of communist eastern Germany to instrumentalize him.

9. Not all of his letters were so courteous. One that he sent from Paris to the American mathematician J. L. Doob was written in a harsh tone. A dry "Monsieur" was all Döblin could muster as salutation before lashing out. Doob had claimed in one of his papers that Döblin and a coauthor had overlooked some "exceptional points" in one of their notes. The young man would have none of that. "When I leafed through it," he wrote, referring to Doob's paper (not "when I read it," that would have been too much honor), "I came across the remark . . . I regret that you neither found it appropriate to send me this paper in which you accuse us of having made an error, nor to write to us before you published this remark. I regret this all the more, since after re-reading the said note I absolutely can't see on what your assertion is based." Doob realized that he had not acted correctly and jotted onto the letter: "Of course, by now I don't have the slightest idea who was right. But of course the complaint is justified, I should have written him."

10. Maybe Frank Ramsey should be included in this list. He died from jaundice in 1930, at age 26, leaving behind a mathematical opus that would prove invaluable to economics.

CHAPTER THIRTEEN

1. Lately a backlash has arisen against the exaggerated use of mathematics in economics. Whether quantification has helped or hurt the discipline is now the subject of debate. Some people argue that a preoccupation with mathematical models has led economists to lose sight of the overall picture. Hence the debates continue.

2. Half a century later, the pendulum would turn again. Harry Markowitz, of whom more is said in the later part of this chapter, remarked about the newfangled theory of econophysics: "If we restrict ourselves to models which can be solved analytically, we will be modeling for our mutual entertainment, not to maximize explanatory or predictive power."

3. It was not love for Jews that had led Schumpeter to defend Samuelson's appointment, as it had not been his disdain for Nazis racism which had spurred his emigration. Actually, he subscribed to distinctly anti-Semitic sentiments himself. Schumpeter considered Jews as different, exclusive, and clannish. They were early bloomers, he believed, who unfairly received more rewards than they would deserve in free competition. His was a mild form of anti-Semitism that was prevalent in the Austro-Hungarian Empire where he grew up, a disease that afflicted nearly every part of society in turn-of-the century Vienna. To his credit, Schumpeter never advocated any kind of discrimination against Jews, as was manifest in Samuelson's case.

4. Samuelson continued his academic work until he died at age 94 in 2009.

5. Harold W. McGraw tells the story of how Basil Dandison, McGraw-Hill's textbook representative, visited Harvard University in search of an author for a textbook on economics. "The day was December 8, 1941, the day after Pearl Harbor. Basil found only one department professor that day, Seymour Harris, who was willing to settle down and discuss a book. Professor Harris mentioned a young man named Samuelson who had come from the University of Chicago and was doing graduate work on the campus. Harris felt that Samuelson would be a good prospect if he would join in an economic symposium and write a couple of chapters for a book to be based on that symposium." Preparing the book took awhile, and Paul Samuelson officially signed on with McGraw-Hill in 1948. The book was revolutionary in more senses than one. As Samuelson proudly pointed out at a celebration on occasion of the book's fiftieth birthday, it pioneered the use of multiple colors in the textbook field.

6. When a paperback edition of *Foundations* was being discussed in 1965, Samuelson hesitated. He feared that "people might be tempted to buy the paperback edition out of miscalculation and error," he wrote in the foreword. What he meant was that students would confuse the rigorous, highly mathematical treatise with the elementary but much more costly ($160 the last time I looked) textbook. His fear was well-founded. When his colleague, the Harvard professor J. Kenneth Galbraith, was asked by a friend to recommend a book typical of what was new in economics, he suggested "you might as well read Samuelson's textbook." As Samuelson tells it in the foreword, the inevitable happened and the next time they met, the friend complained: "The subject of economics has certainly become technical since last I studied it!" Apparently he had mistaken the treatise for the textbook.

7. Nobel Prizes for physics, chemistry, medicine, literature, and peace have been awarded since 1901. In 1969 Sveriges Riksbank, the Swedish central bank, decided to sponsor a prize for economics research in memory of Alfred Nobel. The first laureates were Ragnar Frisch from Norway and Jan Tinbergen from the Netherlands, who were honored "for having developed and applied dynamic models for the analysis of economic processes." The next year's laureate was Paul Samuelson.

8. Important papers on the subject, like "Proof That Properly Anticipated Prices Fluctuate Randomly," "Stochastic Speculative Prices," "General Proof That Diver-

sification Pays," "The Fundamental Approximation Theorem of Portfolio Analysis in Terms of Means, Variances, and Higher Moments," were all published after 1965.

9. How the shape of the utility function explains people's aversion to risk and thus the emergence of the insurance industry is described in Chapter 36 of my book *The Secret Life of Numbers.*

10. No sooner had Osborne's paper appeared in the March-April issue of *Operations Research* than an indignant letter was published, entitled "Comments on 'Brownian Motion in the Stock Market.'" The writer, André G. Laurent of the Department of Mathematics at Wayne State University in Detroit, Michigan, complained that Osborne's main result—that the logarithms of stock prices follow a normal distribution—had been obtained no less than thirteen years earlier. In 1946 in a thesis at the University of Paris, one R. Remery of the French National Institute of Statistics and Economics had analyzed French stock prices since 1929. Laurent conceded that "it is understandable that Mr. Osborne overlooked or was not aware of papers which were published in France or of those which remained unpublished." He was not so forgiving, however, concerning his own work and went on to charge that Osborne should have at least taken note of his, Laurent's, presentation at the annual meeting of the Operations Research Society in Philadelphia in 1957. After all, a thirteen-line abstract of the talk "A Stochastic Model for the Behavior of a Set of Price Index-Numbers and Its Applications," was published in *Operations Research* in the same year. In a touching display of his disappointment with the silence that greeted his work—Osborne's oversight being a case in point—he wistfully points out that the handout he distributed at that meeting "still is often requested."

In a "Reply to 'Comments on Brownian Motion in the Stock Market,'" published right behind the letter, Osborne acknowledges that he had not known about a number of earlier references. He then gives a survey of the literature so that "readers may become better aware than I was at the time of writing, of the history of the ideas on the subject." He points out that lognormal distributions arise in many different contexts, for example, in astronomy, statistical mechanics, biology, and social sciences, and gives due credit to Remery and Laurent for applying them to finance.

11. For the mathematically inclined: log(today's price/initial price) = log(today's price)–log(initial price). Hence, to repeat the above numerical example: if a share is bought for \$100 and sold for \$120, then the profit is 20 percent $(120/100 - 1 = 0.2)$. In terms of logarithms, we have log(120) = 2.079, log(100) = 2. Thus the profit, in terms of logarithms, is 0.079.

CHAPTER FOURTEEN

1. Black's father, an electrical engineer, had worked at a power company in Washington, D.C., edited and then published a trade journal, obtained a law degree, started a brokerage business, and become executive vice president of a power company in Florida. His mother was a homemaker who, with the consent of her husband, made all the big decisions in the house.

2. Herbert Simon won the 1978 Nobel Prize in economics "for his pioneering research into the decision-making process within economic organizations."

3. The model was developed further in the following years by Sharpe, John Lintner, and Jan Mossin.

4. The Dow Jones Index does not quite measure market performance. It considers only the shares of blue-chip companies, does not weigh them by the company size, and does not reinvest dividends that are paid out.

5. Meanwhile, Black's personal life was overshadowed by a tragedy. He had met and fallen in love with a wonderful girl, a tall and slender third grade teacher. The couple had already set a wedding date when fate intervened. Black's fiancée was diagnosed with Lou Gehrig's disease and died in April 1966. He was distraught. With the help of a colleague and friend at ADL he eventually recovered and a year later married Mimi Allen, a psychology student who stunned everybody with her beauty.

6. In 1974 Columbia named him University Professor, the highest academic rank in the university's hierarchy. Merton received a Guggenheim Fellowship, a MacArthur Fellowship, was elected to science academies, and was awarded the U.S. National Medal of Science as well as numerous honorary doctorates from universities around the world. Little did he know that his son would surpass his stellar achievements.

7. The authors were an odd couple. Sheen T. Kassouf had studied mathematics at Columbia University and then ran a successful technical illustration firm. But he was always interested in the stock market. Once, he hesitated whether to invest in a stock or in a warrant and started to examine the relationship between the prices of the two. Eventually he returned to Columbia University to get a Ph.D. in economics, studying warrant pricing. At the same time, he continued trading and managed to double a $100,000 investment in four years. After obtaining his Ph.D., he was appointed professor at the newly established Irvine campus of the University of California.

There he met Edward O. Thorp, a mathematics professor, sometime fund manager, and gambler. Thorp got his Ph.D. in mathematics from UCLA and then worked at MIT for a couple of years. Together with Claude Shannon, the computer pioneer and "father of information theory," Thorp developed a wearable computer at MIT, the first of its kind, in order to play roulette. This enterprise was so successful that the use of computers is now forbidden in casinos. Using probability theory and statistics, Thorp also developed a card counting scheme for blackjack that was so effective that suspicious casino managers in Las Vegas, Reno, and Lake Tahoe had their security people kick him out of their establishments. From 1965 to 1982, he was professor of mathematics and finance at UC Irvine.

CHAPTER FIFTEEN

1. Recall that warrants are a certain kind of option, sold by a corporation. When the time comes, the corporation does not promise to buy a share on the open market

in order to give it to the buyer, but to issue it. That is, additional shares are printed and given to the buyer of the warrant. Thus the holdings of the current stockholders are diluted—the company's assets are now divided by a larger number of shares—but this is not our concern here. In all aspects concerning their value, warrants are identical to options.

2. Meticulous to a fault, Black was known for keeping files on everything he ever wrote or studied. Thus he could claim, twenty years later, that he still possessed the notes that contain the crucial differential equation.

3. Hence the Greeks do not just give the sign of the first derivative but also its numerical value.

4. Since delta is the first derivative, gamma is the second derivative.

5. The correct hedging ratio changes also with the progression of time toward the exercise date.

6. It may not have been the return address. Consider the affiliation that Valery Fabrikant, an applied mathematician and multiple murderer, gives in his numerous publications: Prisoner #167932D, Archambault Jail, Canada.

7. Milton Friedman's parents hailed from Bérégszasz, a small, predominantly Jewish village in Hungary, where they were neighbors of my mother's.

8. Black stayed at Chicago for three years. But his family longed to return to Boston and in 1975, Robert Merton succeeded in luring him to MIT. Scholes moved from Chicago to Stanford University in 1981.

CHAPTER SIXTEEN

1. This investment bank had been founded in 1869 by Marcus Goldman, a Jewish immigrant from Bavaria. Its original name was Marcus Goldman & Company. Goldman made a very successful living, buying up promissory notes from tobacco and diamond dealers in Manhattan and then selling them, with a small markup, to commercial banks. In 1882 his son-in-law Samuel Sachs, himself the son of German Jewish immigrants, joined the firm, which was later renamed Goldman Sachs. The young partner and his close friend Philip Lehman—he too the son of Jewish immigrants from Germany—pioneered the flotation of shares to allow newly founded companies to raise money. It would become Goldman Sachs's core business. By 2010 its total assets had grown to more than $900 billion. By the way, Philip Lehman joined his father Mendel and his uncles Haim and Mayer, collectively known as the Lehman Brothers, in the investment bank of that name. It filed for bankruptcy in September 2008.

2. Salomon Brothers was founded in New York City in 1910 when Arthur, Herbert, and Percy Salomon decided to leave their father Ferdinand's operation. Ferdinand Salomon had immigrated from Alsace-Lorraine, France, and founded a money lending institution in 1880. The three sons joined him in the business but frictions soon developed. As an orthodox Jew, Ferdinand insisted on sanctifying the Sabbath

by foregoing all business dealings on that day. The sons objected. Eventually Arthur, Herbert, and Percy decided to go into business for themselves. After scraping together personal savings of $5,000, they founded Salomon Brothers. Continuing in their father's footsteps, the sons at first continued brokering money but soon branched out to trade corporate bonds and Treasury bills. Salomon Brothers became one of the world's largest investment banks, before it was acquired and ultimately incorporated into Citigroup in 1998.

3. By teaching winter quarters in mild California, Scholes was able to avoid New York's harsh hibernal climate.

CHAPTER SEVENTEEN

1. The trader himself, Paul Mozer, was sentenced to four months in prison and fined $30,000.

2. By the way, they don't give back money when the fund loses.

3. The standard deviation of LTCM's daily profits or losses was $45 million. This gives a 99 percent confidence level that losses in a single day would not exceed $105 million (2.33 times 45 million).

4. Nevertheless, Rosenfeld described the rise from 8 cents to 10 cents, somewhat tongue in cheek, as a 25 percent rise, which sounds much better.

5. Merton had developed the software, called SmartNest, to help individuals plan their pension funds.

BIBLIOGRAPHY

Bachelier, Louis. "Théorie de la Spéculation." *Annales Scientifiques de l.E.N.S* 17 (1900): 21–86.

Barbut, Marc, and Laurent Mazliak. *Commentary on the Notes for Paul Lévy's 1919 Lectures on the Probability Calculus at the École Polytéchnique.* Vol. 4. 2008.

Bernstein, Peter L. *Against the Gods: The Remarkable Story of Risk.* John Wiley, 1996.

———. *Capital Ideas: The Improbable Origins of Modern Wall Street.* Paperback ed. John Wiley, 2005.

———. *Capital Ideas Evolving.* John Wiley, 2007.

Bingham, N. H. "Studies in the History of Probability and Statistics XLVI. Measure into Probability: From Lebesgue to Kolmogorov." *Biometrika* 87 (2000): 145–156.

Black, Fischer. "How We Came Up with the Option Formula." *Journal of Portfolio Management* 15 (1989): 4–8.

Black, Fischer, and Myron Scholes. "The Pricing of Options and Corporate Liabilities." *Journal of Political Economy* 81 (1973): 637–654.

Boness, J. "Elements of a Theory of Stock Options Value." *Journal of Political Economy* 72 (1964): 163–175.

Bony, Jean-Michel, Gustave Choquet, and Gilles Lebeau. "Le centenaire de l'intégrale de Lebesgue." *Comptes Rendus de l'Académie des Sciences de Paris* 332 (2001): 85–90.

Bookstaber, Richard. *A Demon of Our Own Design.* John Wiley, 2007.

———. *Option Pricing and Strategies in Investing.* Addison-Wesley, 1981.

Bru, Bernard. "Le vie et l'oeuvre de W. Doeblin (1915–1940) d'après les archives Parisiennes." *Mathématiques et Sciences Humaines* 119 (1992): 5–51.

———. "Notes de lecture du pli cacheté." *Comptes Rendus de l'Académie des Sciences de Paris* 331 (2000): 1103–1128.

———. "Un hiver en campagne." *Comptes Rendus de l'Académie des Sciences de Paris* 331 (2000): 1033–1035.

Bru, Bernard, and Marc Yor. "Comments on the Life and Mathematical Legacy of Wolfgang Doeblin." *Finance and Stochastics* 6 (2002): 3–47.

Buser, Pierre. "Les plis cachetés du passé aud présent." *La Lettre de l'Académie des Sciences* 2 (2001): 15–17.

Campbell, John Y., Andrew W. Lo, and A. Craig MacKinlay. *The Econometrics of Financial Markets.* Princeton University Press, 1997.

Case, James. "First Gauss Prize Is Awarded to Kiyoshi Itō in Madrid." *SIAM News* 39 (December 2006).

Courtault, Jean-Michel. "Louis Bachelier on the Centenary of *Théorie de la Spéculation.*" *Mathemtical Finance* 10 (2000): 341–353.

Davis, Mark, and Alison Etheridge. *Louis Bachelier's Theory of Speculation: The Origins of Modern Finance.* Princeton University Press, 2006.

De la Vega, Joseph. *Die Verwirrungen der Verwirrungen.* Börsenmedien AG, Kulmbach, 2004. First published in 1688 as *Confusión de confusiones: Diálogos curiosos entre un philosopho agudo, un mercader discreto, y un accionista erudito, descriviendo el negocio de las acciones, su origen, su ethimologia, su realidad, su juego, y su enredo.*

Derman, Emanuel. *My Life as a Quant.* John Wiley, 2004.

———. "Sur l'équation de Kolmogoroff." *Comptes Rendus de l'Académie des Sciences de Paris* 210 (1938): 705–707.

Döblin, Wolfgang. "Sur l'équation de Kolmogoroff." *Comptes Rendus de l'Académie des Sciences de Paris* 331 (2000): 1059–1102.

Edwards, Franklin R. "Hedge Funds and the Collapse of Long-Term Capital Management." *Journal of Economic Perspectives* 13 (1999): 189–210.

Einstein, Albert. "Über die von der molekularkinetischen Theorie der Wärme geforderte Bewegung von in ruhenden Flüssigkeiten suspendierten Teilchen." *Annalen der Physik*, 1905, 549–559.

Freedman, David. *Brownian Motion and Diffusion.* Holden-Day, 1983.

French, Douglas E. *Early Speculative Bubbles and Increases in the Money Supply.* Ludwig von Mises Institute, 2009.

Friedman, Robert Marc. *The Politics of Excellence: Behind the Nobel Prizes in Science.* Henry Holt, 2001

Fulinski, A. "On Marian Smoluchowski's Life and Contributions to Physics." *Acta Physica Polonica B* 29 (1998): 1523–1538.

Galai, Dan. "The Option Pricing Model and the Risk Factor of Stock." *Journal of Financial Economics* 3 (1976): 53–81.

Galignani, A. and W. *Galignani's New Paris Guide, for 1856.* A. and W. Galignani, 1856.

Garber, Peter M. "Tulipmania." *Public Choice* 97 (1989): 535–560.

Girlich, Hans-Joachim. "Bachelier's Predecessors: Address to the 2nd World Congress of the Bachelier Finance Society." Crete, June 2002.

Goede, Marieke de. "Discourse of Scientific Finance and Failure of Long-Term Capital Management." *New Political Economy* 6 (2001): 149–170.

Hafner, Wolfgang. *Im Schatten der Derivate: Das schmutzige Geschäft der Finanzelite mit der Geldwäsche.* Eichborn, 2002.

Hafner, Wolfgang, and Heinz Zimmermann. *Vinzenz Bronzin's Option Pricing Models: Exposition and Appraisal.* Springer-Verlag, 2009.

Hald, A. T. N. "Thiele's Contributions to Statistics." *International Statistical Review* 49 (1981): 1–20.

Harrison, Michael J. *Brownian Motion and Stochastic Flow Systems.* John Wiley, 1985.

Haw, Marc D. "Colloidal Suspension, Brownian Motion, Molecular Reality: A Short History." *Journal of Physics: Condensed Matter* 14 (2002): 7769–7779.

———. "Einstein's Random Walk." *Physics World,* January 2002, 19–22.

Henri, Victor. "Étude cinématographique des mouvements browniens." *Comptes Rendus de l'Académie des Sciences de Paris* 146 (1908): 1024–1025.

Imkeller, Peter, and Sylvie Roelly. "Die Wiederentdeckung eines Mathematikers: Wolfgang Döblin." *MDMV* 15 (2007): 154–159.

Jarrow, Robert, and Philip Protter. "A Short History of Stochastic Integration and Mathematical Finance: The Early Years, 1880–1970." *IMS Lecture Notes Monographs* 45 (2004): 1–17.

Jerisan, David, and Daniel Stroock. "Norbert Wiener." *Notices of the AMS* 42, no. 4 (1995): 430–438.

Jovanovic, Franck. "Éléments biographiques inédits sur Jules Regnault (1834-1894), Inventeur du modèle de marché aléatoire pour représenter les variations boursières." *Revue d'Histoire des Sciences Humaines* 11 (2004): 215–230.

Jovanovic, Franck. "Instruments et théorie économiques dans la construction de la "Science de la Bourse" d'Henri Lefèvre." *Sciences Humains* 2, no. 7 (2002): 41–68.

Kahane, Jean-Pierre. "L'affaire Doeblin: Mon souvenir du pli cacheté de Wolfgang Doeblin." *Mathématiques et Sciences Humaines* 44 (2006): 4–11.

Kassouf, Sheen T., and Edward O. Thorpe. *Beat the Market: A Scientific Stock Market System.* Random House, 1967.

Kauffman, George B. "Review of Marc Friedman's *The Politics of Excellence: Behind the Nobel Prizes in Science.*" *Angewandte Chemie* 42 (2003): 1194–1196.

Kendall, David. "Obituary: Andrei Nokolaevich Kolmogorov (1903–1987)." *Bulletin of the London Mathematical Society* 22 (1990): 31–47.

Kerker, Milton. "The Svedberg and Molecular Reality." *ISIS* 67 (1976): 190–216.

———. "The Svedberg and Molecular Reality: An Autobiographical Postscript." *ISIS* 77 (1986): 278–282.

Kolmogoroff, Andrei. *Grundbegriffe der Wahrscheinlichkeitsrechnung.* Julius Springer, 1933.

———. "Über die analytischen Methoden in der Wahrscheinlichkeitsrechnung." *Mathematische Annalen* 104 (1931): 415–458.

Langevin, Paul. "Sur la théorie du mouvement brownien." *Comptes Rendus de l'Académie des Sciences de Paris* 146 (1908): 530–531.

Lauritzen, Steffen L. "Time Series Analysis in 1880: A Discussion of Contributions Made by T.N. Thiele." *International Statistical Review* 49 (1981): 319–331.

Laurent, A. G. "Comments on Brownian Motion in the Stock Market." *Operations Research* 7 (1959): 806–807.

———. *Thiele: Pioneer in Statistics.* Oxford University Press, 2002.

Lebesgue, Henri. "Sur une généralisation de l'intégrale définie." *Comptes Rendus de l'Académie des Sciences de Paris* 132 (1901): 1025–1028.

Lefèvre, Henri. "Physiologie et méchanique sociales." *Journals des Actuaires Français* 2 (1873): 211–250.

———. "Physiologie et méchanique sociales, 2ème partie." *Journals des Actuaires Français* 2 (1873): 351–388.

———. "Physiologie et méchanique sociales, 3ème partie." *Journals des Actuaires Français* 3 (1874): 93–118.

Lehmann, Bruce N., ed. *The Legacy of Fischer Black.* Oxford University Press, 2005.

Li, T., S. Kheifets, D. Medellin, and M. Raizen. "Measurement of the Instantaneous Velocity of a Brownian Particle." *Science* 328 (2010): 1673–1675.

Lo, Andrew W., and A. Craig MacKinlay. "Stock Prices Do Not Follow a Random Walk." *Review of Financial Studies* 1 (1988): 41–66.

Lowenstein, Roger. *When Genius Failed: The Rise and Fall of Long-Term Capital Management.* Random House, 2000.

Mackay, Charles. *Extraordinary Popular Delusions and the Madness of Crowds.* 1841.

MacKenzie, Donald. "An Equation and Its Worlds." *Social Studies of Science* 33 (2003): 831–868.

Malliaris, A. G. "Itō's Calculus in Financial Decision Making." *SIAM Review* 25 (1983): 481–496.

Mehrling, Perry. *Fischer Black and the Revolutionary Idea of Finance.* John Wiley, 2005.

Merton, Robert C. "Theory of Rational Option Pricing." *Bell Journal of Economics and Management Science* 4 (1973): 141–183.

Neal, Larry. "How It All Began: The Monetary and Financial Architecture of Europe During the First Global Capital Markets, 1648–1815." *Financial History Review* 7 (2000): 117–140.

Neal, Larry, and Lance Davis. "The Evolution of the Rules and Regulations of the First Emeerging Markets: The London, New York, and Paris Stock Exchanges, 1792–1914." *Quarterly Review of Economics and Finance* 45 (2005): 296–311.

Newburgh, Ronald, Joseph Peidle, and Wolfgang Rueckner. "Einstein, Perrin, and the Reality of Atoms: 1905 Revisited." *American Journal of Physics* 74 (2006): 478–481.

Nye, Mary Jo. "Science and Socialism: The Case of Jean Perrin in the Third Republic." *French Historical Studies* 9 (1975): 141–169.

Osborne, M. F. M. "Brownian Motion in the Stock Market." *Operations Research* 7 (1959): 145–173.

———. "Periodic Structure in the Brownian Motion of Stock Prices." *Operations Research* 10 (1962): 345–379.

Parthasarathy, K. R. "Andrei Nikolaevich Kolmogorov." *Journal of Applied Probability* 25 (1988): 445–450.

Perrin, Jean. "L'agitation moléculaire et le mouvement brownien." *Comptes Rendus de l'Académie des Sciences de Paris* 146 (1908): 967–968.

Piasecki, Jaroslaw. "Centenary of Marian Smoluchowski's Theory of Brownian Motion." *Acta Physica Polonica B* 38 (1998): 1623–1630.

Pochet, Léon. "Géométrie des jeux de bourse." *Journals des Actuaires Français* 2 (1873): 153–160.

Poitras, Geoffrey. "The Early History of Option Contracts." In Hafner and Zimmermann, *Vinzenz Bronzin's Option Pricing Models*.

Preda, Alex. "Informative Prices, Rational Investors: The Emergence of the Random Walk Hypothesis and the Nineteenth-Century "Science of Financial Investments." *History of Political Economy* 36 (2004): 351–386.

Renn, Jürgen. "Einstein's Invention of Brownian Motion." *Annals of Physics* 14 (2005): 23–37.

Safranov, Vadim. "The Rise and Fall of the LTCM." Ph.D. seminar. University of St. Gallen, 2000.

Samuelson, Paul Anthony. *Economics: An Introductory Analysis.* 19th ed. McGraw-Hill, 2009.

———. *Foundations of Economic Analysis.* 1947. Reprint. Atheneum, 1965.

———. "Mathematics of Speculative Price." *SIAM Review* 15 (1973): 1–41.

Saunders, E. Stewart. "The Archives of the Académie des Sciences." *French Historical Studies* 10 (1978): 696–702.

Smith, Clifford W. Jr. "Option Pricing." *Journal of Financial Economics* 3 (1976): 3–51.

Smoluchowski, Marian von. "Zur kinethischen Theorie der Brownschen Molekularbewegung und der Suspension." *Annalen der Physik* 21 (1906): 756–780.

Sprenkle, Case. "Warrant Prices as Indicators of Expectations and Preferences." *Yale Economic Essays* 1 (1961): 178–231.

Stonham, Paul. "Too Close to the Hedge: The Case of Long-Term Capital Management LP." *European Management Journal* 17 (1999): pt. 1, 282–289; pt. 2, 382–390.

Szpiro, George G. *Kepler's Conjecture: How Some of the Greatest Minds in History Helped Solve One of the Oldest Math Problems in the World.* John Wiley, 2003.

———. *Poincaré's Prize: The Hundred-Year Quest to Solve One of Math's Greatest Puzzles.* Plume, 2008.

———. *The Secret Life of Numbers: 50 Easy Pieces on How Mathematicians Work and Think.* Joseph Henry, 2006.

————. "Tick Size, the Compass Rose, and Market Nanostructure." *Journal of Banking and Finance* 22 (1998): 1559–1569.

Taqqu, Murad S. "Bachelier et son époque: Une conversation avec Bernard Bru." *Journal de la Société Française de Statistique* 142 (2001).

Thompson, Earl A. "The Tulipmania: Fact or Artifact." *Public Choice* 130 (2006): 99–114.

Velde, François R. *Government Equity and Money: John Law's System in 1720 France.* Federal Reserve Bank of Chicago, 2004.

Viswanathan, G. M., et al. "Lévy Flight Search Patterns of Wandering Albatrosses." *Nature* 381 (1996): 413–415.

Vovk, V. G., and G. R. Shafer. "Kolmogorov's Contributions to the Foundations of Probability." *Problems of Information Transmission* 39 (2003): 21–31.

Weber, Ernst Juerg. *A Short History of Derivative Security Markets.* 2008. http://ssrn.com/abstract=1141689.

Whelan, S. F., D. C. Bowie, and A. J. Hibbert. "A Primer in Financial Economics." *British Actuarial Journal* 8 (2002): 27–74.

White, Eugene N. *The Crash of 1882, Counterparty Risk, and the Bailout of the Paris Bourse.* NBER Working Paper Series no. 12933. 2007.

Wiener, Norbert. *I Am a Mathematician: The Later Life of a Prodigy.* Doubleday, 1956.

Willis, H. P., and G. Francois. "The Paris Bourse." *Journal of Political Economy* 6 (1898): 536–544.

Wiston-Glynn, A. W. *John Law of Lauriston: Financier and Statesman.* E. Saunders, 1907.

Yor, Marc. "Présentation du pli cacheté." *Comptes Rendus de l'Académie des Sciences de Paris* 331 (2000): 1037–1058.

Zimmermann, Heinz, and Wolfgang Hafner. "Amazing Discovery: Vincenz Bronzin's Option Pricing Model." *Journal of Banking and Finance* 31 (2007): 532–546.

Zola, Emile. *L'Argent.* 1891. Reprint. Gallimard, 1972.

INDEX